ESSENER GEOGRAPHISCHE ARBEITEN 37

ANJA SCHEFFERS

COASTAL RESPONSE TO EXTREME WAVE EVENTS

HURRICANES AND TSUNAMI ON BONAIRE

A Contribution to IGCP 495

37 ESSENER GEOGRAPHISCHE ARBEITEN

Herausgeber:

Winfried Flüchter, Dieter Kelletat, Wilhelm Kuttler, Ulrich Schreiber und Hans-Werner Wehling

Institut für Geographie, Universität Duisburg-Essen

Schriftleitung: Gudrun Reichert

Die Deutsche Bibliothek - CIP-Einheitsaufnahme

Anja Scheffers

Coastal Response to Extreme Wave Events

Hurricanes and Tsunami on Bonaire – 1. Auflage – Essen: Selbstverlag, 2005
(Essener Geographische Arbeiten, Band 37)
ISBN 3-9808567-3-9

1. Auflage Oktober 2005
© Institut für Geographie, Universität Duisburg-Essen 2005
Druck: Schröers-Druck GmbH, Essen
Alle Rechte vorbehalten

ISBN 3-9808567-3-9

Preface

The horrific Indian Ocean tsunami of 26 December 2004 strengthened warnings that tsunami scientists from all over the world have highlighted often: our written and instrumental records rarely span enough time to warn of the full range of a region's tsunami hazards (ATWATER et al., 2005). Since years our research group studies paleotsunami impacts, yet no one of us could imagine that only weeks after the announcement of the *"First International Tsunami Field Symposium"* as one of the activities of IGCP 495 and CCS of IGU, the most deadly tsunami in human history would occur.

The reason to prepare a volume on extreme wave imprints on the coastlines of Bonaire during the Holocene is manifold:

Since many years, people try to establish a basis for a Caribbean wide tsunami warning system, and a Steering Committee for this purpose under the presidentship of George Maul has done a lot of groundwork. Several meetings – the last two under the mentorship of Aurelio Mercado from Puerto Rico University in March, 2004, and by by Joanne Bourgeois, Harry Yeh and Brian Atwater in Seattle in June, 2005 – have taken place presenting the progress in tsunami research. The necessity of such a task has been clearly demonstrated by the most deadly tsunami in mankind's history in southern Asia on December, 26th, 2004, when ca. 300,000 people died by a tsunami reaching from Indonesia to eastern Africa. This tragedy incorporates the chance of establishing warning systems and mitigation programs in respect to these forces of nature.

Secondly, the ongoing discussion on the capability of hurricane and tsunami waves to break off coastal rock and transport large boulders, important processes in coastal geomorphology, provided a point of controversy we wanted to address. Physical equations on boulder transport and peer-reviewed tsunamigenic boulder transport publications exist together with direct observations on the power of storms, and a lot of arguments for tsunami moved boulders of extreme size have been published. Nevertheless, most of the coastal scientists raise doubt of tsunami induced boulder transport. To our knowledge, the island of Bonaire represents the most favourable place to demonstrate the different sedimentologic imprints of hurricane and tsunami waves on rocky shorelines.

The International Geoscience Correlation Program 495 "Quaternary Land-Ocean Interactions: Driving Mechanisms and Coastal Responses" under the chairmanship of Antony Long (Durham, UK), has established several working groups to adress scientific aims and solve open questions on coastal processes, one of the programs is focusing on tsunami. Therefore this volume is a contribution to IGCP 495, as well. The Commission on Coastal Systems of the International Geographical Union (chairman Doug Sherman, Texas University, USA) has also shown its interest in tsunami research, incorporating several contributions during the last international meeting in Glasgow, Scotland, 2004.

Our research on Bonaire, Aruba and Curaçao started in the year 2000, when for the first time the idea of Holocene tsunami impacts on the neighboring islands of Aruba and Curaçao was addressed. Efforts to determine the source region of at least three strong Holocene tsunami, dated to about 500 BP, 1500 BP and 300 BP, have widened our paleotsunami research to the Antillean Island Arc including Barbados, Grenada, St. Lucia, Guadeloupe, St. Martin and Anguilla, and to the Bahamas in the north. However, the sedimentologic paleotsunami record could extend the historical tsunami catalogue of the Caribbean by LANDER & WHITESIDE (1997), or LANDER, WHITESIDE & LOCKRIDGE (2002), which list 88 tsunami since the age of discovery, far back into the Middle Holocene.

This volume has been written as a guide for the *First International Tsunami Field Symposium* on Bonaire in March, 2006. Short chapters describe the geodynamical development of the Caribbean Plate, Bonaire's geology and geography, and the coral reefs of this part of the world. The main contents discuss the hurricane and tsunami history of Bonaire, presenting field documents and personal observations from the last years including a dating program with ESR. This material forms the base for the field trips. This volume also contains the itinerary for three excursions along nearly all the coastlines of Bonaire as planned for the 2006 Symposium.

Acknowledgements

We are grateful to many people who contributed to our research and the realization of the Symposium:

- STINAPA, Bonaire
- DROB, Sekshon di Medio Ambiente I Naturalesa
- Deutsche Forschungsgemeinschaft for financial support of our research
- Sea Grant Program for financial support of the Symposium
- Institute of Geography, University of Duisburg-Essen, for printing of the field guide
- Zeitschrift für Geomorphologie for printing of the Symposium proceedings
- Technical support from Gudrun Reichert and Anne Hager
- Prof. Dr. Dieter Kelletat and Dr. Sander Scheffers for scientific and personell teamwork and support

And a special thank to all the warmhearted local people of the ABC islands, who helped whenever assistance or support was needed.

We hope that this volume and the Field Symposium will contribute to a better understanding of tsunami hazards worldwide and the Caribbean in particular.

And with the words of Keith Alverson from the Global Ocean Observing System of the Intergovernmental Oceanographic Commission of UNESCO *"Let us hope that we are now taking the first step to ensure that the next tsunami – wherever and whenever it inevitably occurs – will not go down in history as a catastrophe, but as a tribute to the ability of science and technology to serve society"* (ALVERSON, 2005).

We dedicate this volume to the memory of all tsunami victims in the world.

Essen, November 2005

Anja Scheffers

Vorwort

Der katastrophale Tsunami im Indischen Ozean am 26. Dezember 2004 hat wieder einmal belegt, was Wissenschaftler bereits seit längerer Zeit warnend verkündet haben: Direkte Messungen und Beobachtungen von Tsunami und die Zeitspanne schriftlicher Aufzeichnungen ist viel zu kurz, um einen Eindruck vom wirklichen Gefahrenpotential durch Tsunami in einer Region zu gewinnen (ATWATER et al., 2005). Obwohl unsere Arbeitsgruppe seit einigen Jahren die Spuren von Paläo-Tsunami erforscht, hat doch niemand von uns damit gerechnet, dass nur einige Wochen nach der Ankündigung des „First International Tsunami Field Symposium" für den März 2006 als eine der Aktivitäten des IGCP 495 und der Commission on Coastal Systems der Internationalen Geographenunion der tödlichste Tsunami in der Menschheitsgeschichte passieren würde.

Der Grund zur Verfassung dieses Bandes über die Einflüsse extremer Wellenereignisse auf die Küsten von Bonaire im Holozän ist ein mehrfacher:

Seit etlichen Jahren bemüht sich eine Arbeitsgruppe unter der Führung von George Maul um die Etablierung eines Tsunami-Warnsystems für die gesamte Karibik. Wichtige Grundlagen dafür sind bereits erarbeitet. Zur Präsentation der Fortschritte in der Tsunamiforschung haben verschiedene Meetings stattgefunden, die letzten beiden unter der Leitung von Aurelio Mercado aus Puerto Rico im März 2004 auf dieser Insel sowie in Seattle im Juni 2005, organisiert von Joanne Bourgeois, Harry Yeh und Brian Atwater. Die Notwendigkeit, sich weiter intensiv mit den Tsunamigefahren zu beschäftigen wurde durch den tödlichsten Tsunami in der Menschheitsgeschichte in Südostasien am 26. Dezember 2004 belegt, welcher von Indonesien bis Ostafrika seine zerstörerischen Spuren hinterließ. Diese Tragödie beinhaltet aber auch die verstärkte Chance zur Errichtung von Tsunami-Warnsystemen und Schutzprogrammen gegenüber dieser einzigartigen Naturgefahr.

Des Weiteren wollten wir einen Beitrag zu der immer noch diskutierten Frage leisten, inwieweit die Wellen von Hurrikanen und Tsunami in der Lage sind, große Blöcke aus Kliffen herauszubrechen und landwärts zu transportieren. Dazu liegen Veröffentlichungen mit Blockkartierungen, physikalische Berechnungen der zum Blocktransport benötigten Energie und direkte Beobachtungen von Sturmwellen-Kapazitäten zum Blocktransport vor und damit eine ganze Reihe stichhaltiger Argumente, die belegen, dass zum Transport sehr großer Blöcke gegen die Schwerkraft ausschließlich starke Tsunami in der Lage sind. Nichtsdestoweniger bezweifeln viele Küstenforscher immer noch diese Tatsache. Da nach unserer Kenntnis die Insel Bonaire die besten Belege für die Transportvorgänge von großen Blöcken durch Stürme und Tsunami im Vergleich aufweist, haben wir diese Region zur Demonstration dieser Sachverhalte ausgewählt.

Das International Geoscience Correlation Program IGCP 495 "Quaternary Land-Ocean Interactions: Driving Mechanisms and Coastal Responses" unter dem internationalen Leiter Antony Long aus Durham (UK) hat mehrere Arbeitsgruppen mit definierten wissenschaftlichen Zielen zur Küstenforschung gebildet, darunter eine zu Tsunami. Daher ist dieser Band auch ein Beitrag zu IGCP 495. Die Commission on Coastal Systems der Internationalen Geographenunion unter Leitung von Doug Sherman (Texas University, USA) hat ebenfalls ihr Interesse an Tsunamiforschung bekundet, u.a. durch Einbeziehung mehrerer einschlägiger Beiträge in die Sitzungen des letzten Internationalen Meetings in Glasgow (Schottland) im Jahre 2004.

Unsere Feldforschungen auf Bonaire, Aruba und Curacao begannen im Jahre 2000, als zum ersten Mal die Idee von holozänen Tsunami-Wirkungen in diesem Raum der Karibik entstand. Durch den Versuch, die Herkunftsgebiete der auf etwa 500 BP, 1500 BP und 3500 BP datierten Tsunami zu ermitteln, wurden unserer Forschungen an Paläo-Tsunami ausgeweitet auf den Inselbogen der Antillen mit Barbados, Grenada, St. Lucia, Guadeloupe, St. Martin und Anguilla und schließlich auch auf die nördlich gelegenen Bahamas. Dadurch konnte der historische Katalog über Tsunami der lezten 500 Jahre, erstellt von LANDER & WHITESIDE (1997) und LANDER, WHITESIDE & LOCKRIDGE (2002) mit 88 aufgelisteten Tsunami räumlich und zeitlich erheblich erweitert werden bis weit zurück in das mittlere Holozän.

Dieser Band dient als Enführung zum „First International Tsunami Field Symposium" auf Bonaire im März 2006. Kurze Kapitel führen ein in die geodynamische Situation der Karibik, Bonaire's Geologie und Geographie sowie die Korallenriffe in diesem Teil der Erde. Der Hauptteil konzentriert sich auf die erkennbare Geschichte von Hurrikan- und Tsunami-Einflüssen und legt dazu Felddokumente, absolute Datierungen mit Radiokohlenstoff und ESR und eigene Beobachtungen an Hurrikanen vor. Dieses Material bildet gleichzeitig die Grundlage für die Symposiums-Exkursionen und enthält deren Ablauf mit Haltepunkten und Demonstrationsobjekten.

Danksagungen

Wir danken vielen Menschen und Organisationen, die unserer Forschungen unterstützt und damit die Grundlage für dieses Symposium gelegt haben:

- STINAPA, Bonaire
- DROB, Sekshon di Medio Ambiente I Naturalesa
- Deutsche Forschungsgemeinschaft für finanzielle Förderung der Feldarbeiten
- Sea Grant Program der National Science Foundation der USA über die University of Puerto Rico/Prof. Dr. Aurelio Mercado
- Universität Duisburg-Essen für die Unterstützung beim Druck dieses Bandes und der Vorbereitung des Symposiums
- Zeitschrift für Geomorphologie für die Zusage zum Druck der Symposiums-Proceedings
- Gudrun Reichert und Anne Hager für vielfältige technische Unterstützung bei der Erstellung von Abbildungen und Text
- Prof. Dr. Dieter Kelletat und Dr. Sander Scheffers für wissenschaftliche und persönliche Zusammenarbeit und Unterstützung

Ein besonderer Dank geht an die warmherzigen Bewohner der ABC-Inseln, die mit freundlichem Entgegenkommen unsere Arbeiten wo immer möglich gefördert haben.

Wir hoffen, dass dieser Band und das Symposium dazu beitragen werden, Tsunamigefahren weltweit und speziell in der Karibik besser zu verstehen. Um mit den Worten von KEITH ALVERSON vom *Global Ocean Observing System* der *Intergovernmental Oceanographic Commission* der UNESCO zu sprechen:

"Let us hope that we are now taking the first step to ensure that the next tsunami – wherever and whenever it inevitably occurs – will not go down in history as a catastrophe, but as a tribute to the ability of science and technology to serve society" (ALVERSON, 2005).

Wir widmen diesen Band dem Gedenken an alle Tsunamiopfer auf der Welt.

Essen, November 2005

Anja Scheffers

ESSENER GEOGRAPHISCHE ARBEITEN   Band 37   Seite 1-98   Essen 2005

Coastal Response to Extreme Wave Events – Hurricanes and Tsunami on Bonaire

ANJA SCHEFFERS[1]

CONTENTS

Preface 1

Acknowledgements 2

Vorwort 3

Danksagungen 4

Contents 5

1. Introduction: Environment of the Caribbean 7
   1.1 Plate Tectonics and Volcanism 7
   1.2 Geology and Tectonics 8
   1.3 Caribbean Coral Reefs in General and Important Species 8
       1.3.1 Caribbean Reef Morphology 9
       1.3.2 Caribbean Reef Zonation 11
   1.4 Climate, Storms and Hurricanes 13
   1.5 Tsunami in the Intra-Americas Seas 15

2. Geology and Geography of Bonaire 16
   2.1 Pre-Quaternary Basement 16
   2.2 Pleistocene Reefs 17
   2.3 Geomorphology in General 20
   2.4 Coastal Forms and Processes 21
   2.5 Holocene Coral Reefs 26
   2.6 Soil and Vegetation 27
   2.7 Landuse, Infrastructure and Man Made Environmental Impacts 28

3. Storms and Hurricane History of Bonaire 29
   3.1 The Historical Perspective: Storm and Hurricane Tracks of the Last 150 Years 29
   3.2. Hurricane *IVAN* of September, 2004 30
   3.3 Hurricane *LENNY* of November, 1999 41
   3.4 Event of 1877 44
   3.5 Older Storm Relics 49
   3.6. Hurricane Impacts on the Reefs of Bonaire 52

---

[1] Anja Scheffers, Institute of Geography, Faculty of Biology and Geography, University of Duisburg-Essen, Universitätsstr. 15, D-45117 Essen, Germany, email: anja.scheffers@uni-essen.de

4. Tsunami Imprints on Bonaire 54
   4.1 Historical Sources 54
   4.2 Tsunami Sediments of Bonaire 54
   4.3 Relative Dating of Tsunami Events 68
   4.4 Absolute Dating of Tsunami Events and the Tsunami Risk of Bonaire 70
   4.5 Reconstruction of Holocene Reef Communities from Tsunami Deposits 80
   4.6 Comparison of the Bonaire-Results with Curaçao and Aruba 80
   4.7 Comparison of the ABC-Results with the Wider Caribbean 80

5. Discussion: The Role of High Magnitude Events on the Coastal Geomorphology and Sedimentology of Bonaire, and Protection Measures 82

6. Conclusions 82

7. Summary 82

8. Field Trip Itinerary and Facts 84
   8.1 First day, March, 2nd, 2006 84
   8.2 Second day, March 3rd, 2006 84
   8.3 Third day, March 4th, 2006 86

References 87

List of Figures 93

List of Tables 96

Publications – Institut of Geography 98

1. Introduction: Environment of the Caribbean

1.1 Plate Tectonics and Volcanism

A summary of the geodynamical situation of the Caribbean has been published by SCHELLMANN, RADTKE & WHELAN (2004; see also AAPG, 2003; BLUME, 1974; DENGO & CASE, 1990; MACDONALD et al., 2000; PARARAS-CARAYANNIS, 2004; PINDALL & BARRETT, 1990; SIGURDSSON & CAREY, 1991; TOMBLIN, 1975; or WEYL, 1966).

The Caribbean is nearly identical with a lithosphere plate reaching from the Middle American land bridge (closed after 3 Ma ago) to the Antillean island arc in the east (Fig. 1). The Caribbean plate forms part of a region of great geologic and geographic diversity and geologically, has undergone many changes during its complex evolution (PARARAS-CARAYANNIS, 2004; AAPG, 2003). This is the result of a spreading process between North and South America and may have its origin near the Galapagos islands, probably in the Jurassic period. During the older Tertiary a first volcanic chain developed at the eastern end along the Atlantic subduction zone, which now is drowned and inactive:

During Miocene times, a younger and still active volcanic belt of about 800 km of length developed as an inner chain farther to the west. Here the plate subdues under the North American Plate with its Atlantic Section. The southern boundary transform fault cuts the northern coasts of South America, and the northern one ends in the Puerto Rican trench and outside north to the Larger Antilles. The subduction under the Atlantic Ocean bottom with

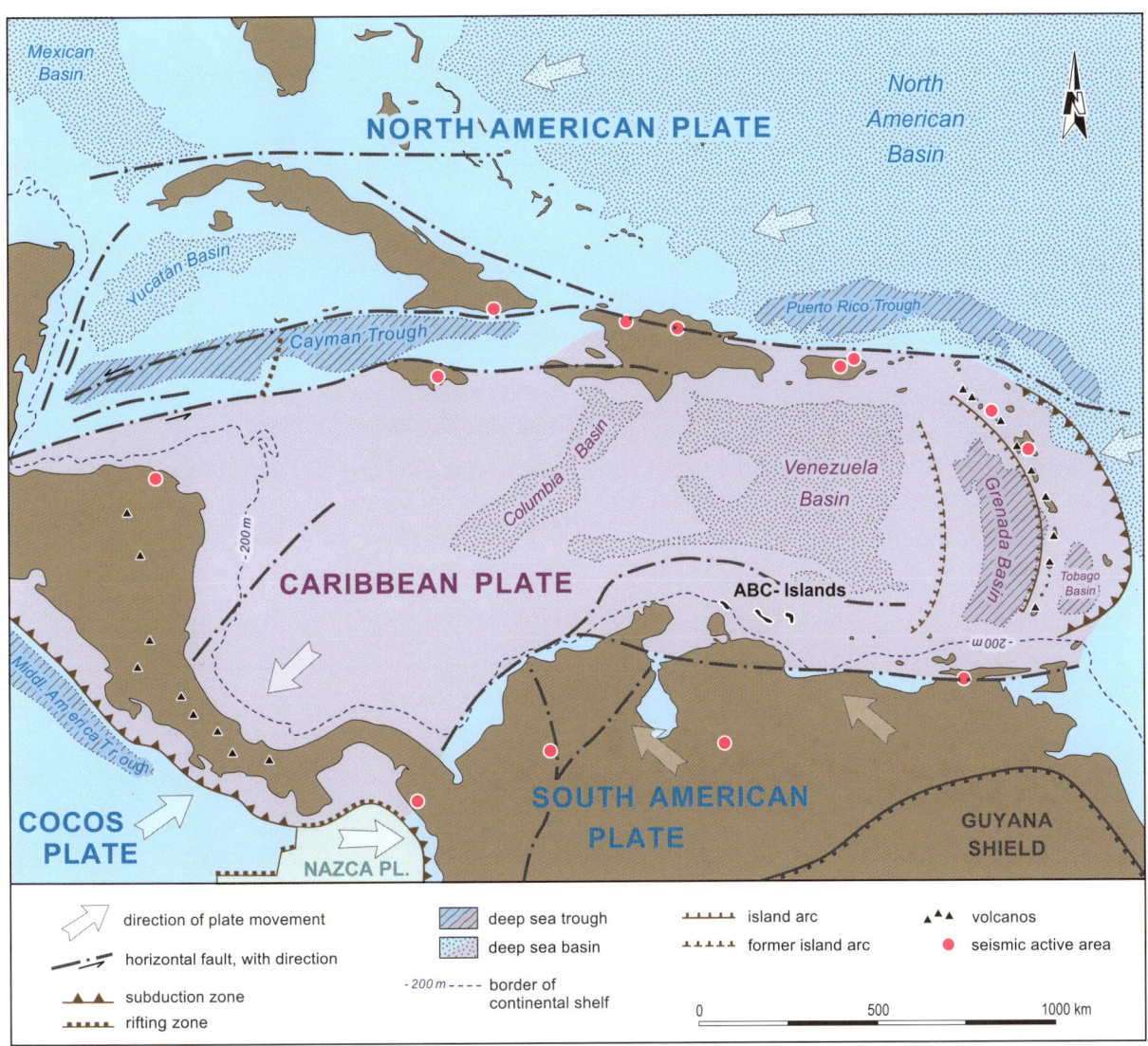

Fig. 1: Geodynamics of the Caribbean (SCHEFFERS, 2002a, modified from SCHUBERT, 1988; MANN et al., 1990; and others).

about 2 cm per year at an angle of 50-60 degrees triggers earthquakes and a chain of active volcanoes, from Montserrat in the north to Kick 'em Jenny, a submarine volcano rising 12 km NW of Grenada close to sea level. The eastern part of the Caribbean plate shows an older sunken arc with only Barbados island as a remnant of its ridge, and the basin west of the arc is divided by the Aves ridge into two parts of different size, the Grenada Basin and the Venezuela Basin. The steep strato-volcanoes of the Antillean arc are of explosive character with major eruptions like of Mt. Pelée on Martinique in 1902, killing most of the 30,000 inhabitants, or the Souffrière volcano on Montserrat with evacuation of many people from that island only years ago. Geological investigations revealed a long history of eruptions during Holocene times, of which many should have induced local destructive tsunamis, but we are still in the beginning of understanding their history from field evidence (PARARAS-CARAYANNIS, 2004).

Additionally, submarine debris avalanches on the sea floor around many islands in the Lesser Antilles suggest that large scale landslides and volcanic island flank collapses must have generated tsunami in the distant past (DEPLUS et al., 2001; PARARAS-CARAYANNIS, 2004). Younger Pleistocene Tsunami deposits with large boulders can be seen on St. Lucia up to at least +50 m (SCHEFFERS et al., 2005a), and on Grenada close to sea level along the west coast.

The ABC-islands Aruba, Bonaire and Curaçao are located on the southern rim of the Caribbean plate, with Aruba on the Venezuelan shelf, and Curaçao and Bonaire separated by 2,000 m deep water. Together they form an interrupted ridge of Tertiary submarine volcanoes, uplifted and covered by Plio-Pleistocene reef limestone.

1.2 Geology and Tectonics

We are mostly interested in the young tectonic history as well as in the petrography, which influences many of the coastal aspects of the region. In general limestone is widespread on all Caribbean islands except of the southernmost ones on the Antillean island arc, which are almost entirely made up of young volcanic rocks, most of them pyroclastics.

Here, vertical movement is difficult to measure because of destructions of former volcanic edifices. This is facilitated at the limestone islands (Barbados, see SCHELLMANN, RADTKE & WHELAN, 2003), Guadeloupe, Anguilla or Puerto Rico as well as Aruba, Curaçao and Bonaire in the south, where limestone deposition started in the Tertiary, mostly as coral debris, and resulted in fringing reefs all around the islands in Pleistocene times.

As will be shown later, these coral limestones and reefs document a long history of uplift and tilting, with some debatable exceptions during the Holocene.

1.3 Caribbean Coral Reefs in General and Important Species

Coral reefs are the most productive (SOROKIN 1993) and diverse (WOOD 2001) ecosystems of the oceans and the largest biological structures on earth. They are generally prominent three-dimensional structures on the sea bottom and are characterized by their growth towards sea level and by their lateral seaward extension.

Coral reefs may be differentiated into several geomorphic units, defined by their shape, size and location relative to the coastline. These units inclu-

Fig. 2: *Millepora* spec.

de fringing reefs, barrier reefs, atolls and patch reefs. Classification based on their position relative to sea level includes drowned reefs, emerged reefs and shallow-water reefs. The prevailing wind direction differentiates reefs in windward and leeward reefs. The number of parallel reef tracts determines the type of reef: single, double or multiple reefs.

The assessment of the relationship between sea level rise and reef response allows for the differentiation between (SPENCER & VILES 2002, NEUMANN & MCINTYRE 1985):

1. "Give-up" reefs, which are relics drowned by sea level rise, non-carbonate sedimentation etc.,
2. "Catch-up" reefs, which fast sea level rise initially left behind, but which recovered with accelerated vertical reef growth during times of slower sea level rise,
3. "Keep-up" reefs, which are able to maintain their reef crest at or near the rising sea level.

1.3.1 Caribbean Reef Morphology

The morphologic terminology developed for Discovery Bay, Jamaica by GOREAU & LAND (1974) is often referred to as "typical" for Caribbean reefs. While variations occur from place to place, this general scheme is a reasonable place to start, and each element of this profile is discussed below.

Shallow Reef

The crest of the main reef is generally emergent in Pacific reefs at low tide, but may be below the surface in Caribbean reefs. The seaward edge of the reef crest takes the brunt of the incoming wave energy. ROBERTS (1989) has shown that the reef can reduce incoming wave energy by up to 97%. As waves break, water is washed across the reef crest and into the lagoon, driving lagoonal circulation (HUBBARD et al., 1981).

Because of the modification of wave forces across the reef crest, the backreef is an environment of totally different physical processes, ecology and sediment characteristics. Sediments and rubble from the reef crest are dumped behind the crest, widening the backreef flat through time. Caribbean reef flats are relatively narrow compared to other places in the world. While zonation is less pronounced, there is a general transition from branching corals and the hydrozoan *Millepora* spec. (Fig. 2) near the front of the crest to sand flats and seagrass landward. The shallow back reef may have a shallow *Porites* reef flat (Fig. 3) immediately behind the crest and numerous small patch reefs surrounded by sand. The corals are generally well adapted to the high levels of sedimentation to which they are regularly subjected. In the Caribbean, the dominant corals include *Porites porites*, *Madracis mirabilis* (Fig. 4) and several head corals, especially the *Montastraea annularis* complex (Fig. 5), *Porites asteroides* and species of *Diploria* (Fig. 6a, b).

Forereef

The forereef extends seaward and downward from the reef crest. It is the most complex of the reef zones, owing to the large depth gradient over which it occurs. In many areas, the forereef is organized into a set of en-echelon reef promontories and sand channels, termed "spur-and-groove" topography. Spur-and-groove is common in both modern and ancient reefs. The term was originally coined from Indo-Pacific examples formed by erosion of the algal rim just below the surf zone. More recently, examples have

Fig. 3: *Porites porites*

Fig. 4: *Madracis mirabilis*

Fig. 5: *Montastrea annularis*

Fig. 6: (a) *Diploria labyrinthiformis*, (b) *Diploria strigosa*

been described from the Caribbean that appears to be the result of accretion by *Acropora palmata* under the influence of strong wave surge.

Both the coral branches and the intervening sand channels are oriented parallel to the dominant wave-approach direction. SHINN (1963) and ROBERTS (1974) proposed that the channels serve as primary conduits for sediment export from the reef. They further proposed that spur-and-groove topography will be best developed along windward margins where a barrier exists to bankward transport, and down slope sediment movement is the only means of export.

Forereef Slope and Deep Forereef

The forereef slope is the least consistent of any of the reef zones, in either its occurrence or character. At many sites, it is totally absent and the forereef drops from shallow water to oceanic depths. Where a forereef slope is present, the deep forereef usually occurs as a well-defined ridge near the platform margin.

Otherwise, it is simply a down-dip extension of the forereef. When occurring separate from the shallower reef zones, the location of the deep forereef is probably controlled by both the break in slope and the existence of an antecedent high left by a previous reef. The character of the reef surface is often similar to the spur-and-groove topography described above, except that the scale of both the reef promontories and the intervening channels is generally larger.

The Reef Wall

Perhaps the most dramatic feature of the deep forereef is the "reef wall". At depths ranging from 50 to 85 meters around the Caribbean, the forereef slope rolls over to a vertical or, in some places, overhanging precipice. The role of active accretion at this depth is not well understood, owing to its remoteness.

1.3.2 Caribbean Reef Zonation

Early discussions of reef zonation were based on Caribbean reefs and, therefore, reflect their species composition, but a profile across a "typical" Caribbean reef shows both morphological and species zonation. The following is a generalized description of species patterns keyed to that profile. Though, it should be kept in mind that the distribution of reef inhabitants could vary from this general picture in response to local physical-oceanographic processes.

The *Acropora palmata* Zone

Along the front of moderate- to high-energy reefs, the dominance of *Acropora palmata* (Fig. 7) to depths of 5-10 m is primarily a response to wave energy. While, not as strong as head corals, *Acropora palmata* orients its branches to minimize tensional loading – the higher the current flow, the lower is the profile presented by the branches of the colony. *Acropora palmata* is capable of rapid growth (>20 cm y^{-1}), which elevates the colony above the traction carpet of shifting sediment that can "sandblast" polyps sitting closer to the bed. Because *Acropora palmata* is extremely sensitive to sedimentation and has no physiological mechanism for sediment removal, wave-induced surge performs this important function for the coral.

The Massive Coral Zone

Usually starting at depths of 5-10 m, head corals increase in importance; notable among these are *Montastraea annularis*, *Montastraea cavernosa* (Fig. 8), *Porites asteroides*, *Colpophylla natans* (Fig. 9) and species of *Diploria*. In some areas, the branching coral *Acropora cervicornis* (Fig. 10) can occur as a narrow belt between this and the shallower *Acropora palmata* zone. Both their depth of occurrence and their massive nature make head corals in this zone more resistant to periodic disturbance by storms. As a result, this zone is often the best represented in cores from Holocene reefs throughout the Caribbean region.

The Platy Coral Zone

At depth, most corals are platelike, an adaptation that concentrates the photo-receptive algae contained within their tissues along the upper surface of the colony. In shallow water, this fragile shape is disadvantageous from a structural point of view. In deeper water, however, physical breakage is low and the need to gather light dominates. The primary species at depth vary from place to place, but

Fig. 7: *Acropora palmata*

Fig. 8: *Montastrea cavernosa*

Fig. 9: *Colpophylla natans*

Fig. 10: *Acropora cervicornis*

usually include members of the genus *Agaricia* (Fig. 11) and flattened colonies of *Montastrea annularis*. As said, this rudimentary classification of coral reef zones is strongly modified by local ecological conditions. For example, wave strength is one important environmental factor for the distribution of coral species. Wave action, on the other hand, depends on reef morphology and varies between forereef, forereef slope, and the deep zone. This leads to wave dependent changes in reef zones.

Fig. 11: *Agaricia* spec.

1.4 Climate, Storms and Hurricanes

The climate of the Caribbean is tropical warm with average annual temperatures of 26°C to 28°C. Rainfall is mostly concentrated in the last quarter of the year with amounts of 450 to >1,000 mm, depending on relief and exposure to trade winds and easterly storms and hurricanes. Wind is very persistent throughout the year and in the case of Bonaire a trade wind blows nearly directly from the east, in the sector 80 to 85 degrees. The trade wind is rather strong and able to shape all vegetation (e.g. Divi Divi trees). Its average speed is around 25 km/h. Water temperature on average is around 27°C.

The Caribbean lies in the tracks of topical storms and hurricanes from the open Atlantic with the exception of the southernmost islands like Aruba, Curaçao and Bonaire, which are visited only very rarely by these storms (e.g. Holland, 1997; KELLE-TAT & SCHEFFERS, 2001; SCHEFFERS, 2002a; LUDLUM, 1989; METEOROLOGICAL SERVICE, 2002; MILLAS, 1968; NEUMANN et al., 1993; RAPPAPORT & FERNANDEZ-PARTAGAS, 1997; READING, 1990; UNISYS CORPORATION, 2002).

Nevertheless they may leave significant imprints on coastal forms and sediments (see chapters 3.1 to 3.5). During the last hundred years approximately 1000 tropical storms and about 200 hurricanes of category 2 to 5 (i.e. sustained winds of >170 km/h) occurred in the Intra Americas Seas region (Fig. 12). In 2004 alone five hurricanes exhibited enormous destructive power on many shorelines in the Caribbean. According to READING (1990), the ABC islands are situated in quadrants of less storm activities, although they experienced two strong hurricanes during the last seven years (*LENNY*, 1999, and *IVAN*, 2004). Besides the destruction by wind forces and inundation by storm surge and wave at-

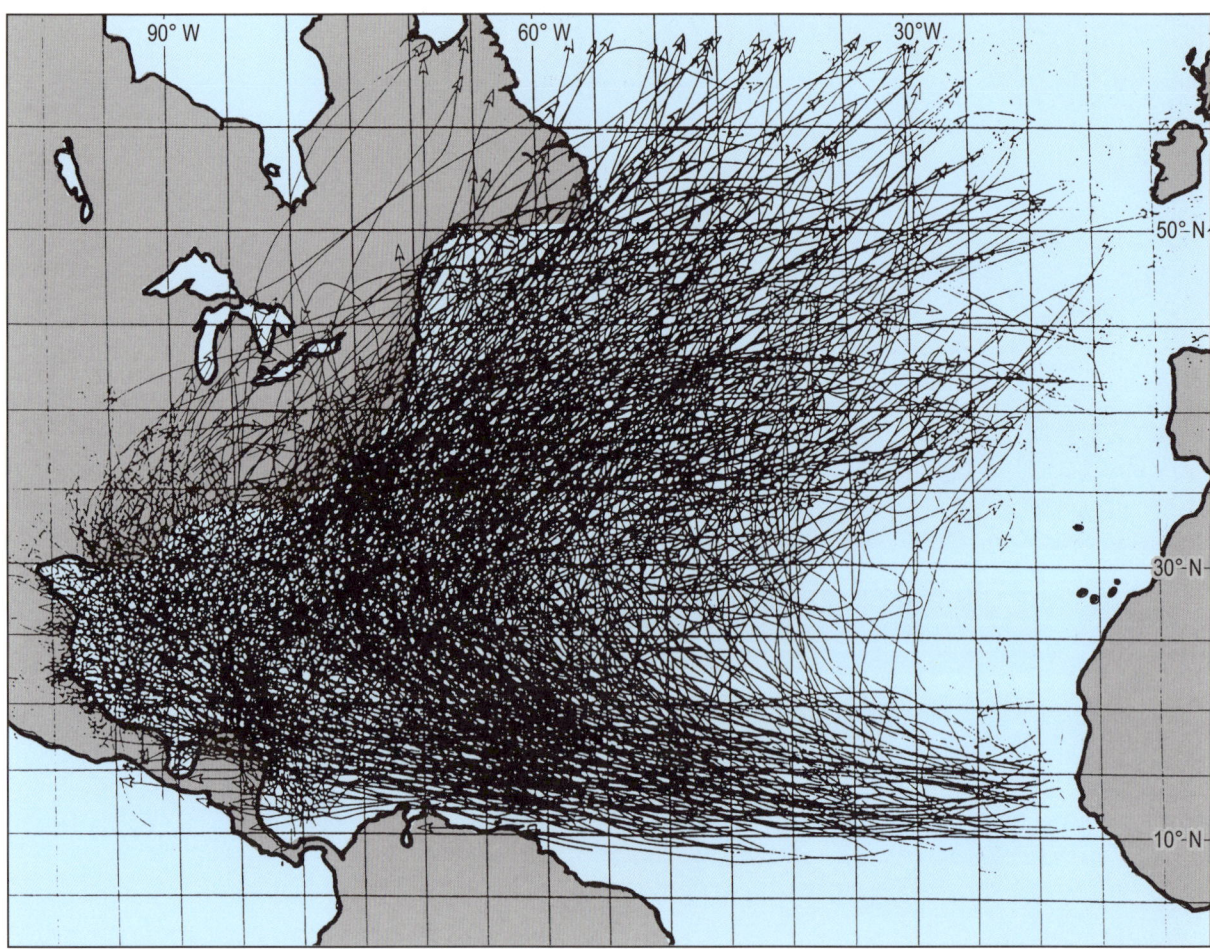

Fig. 12: Storm tracks in the Intra Americas Seas from 1886 to 1995 (METEOROLOGICAL SERVICE OF THE NETHERLANDS ANTILLES AND ARUBA, 2002).

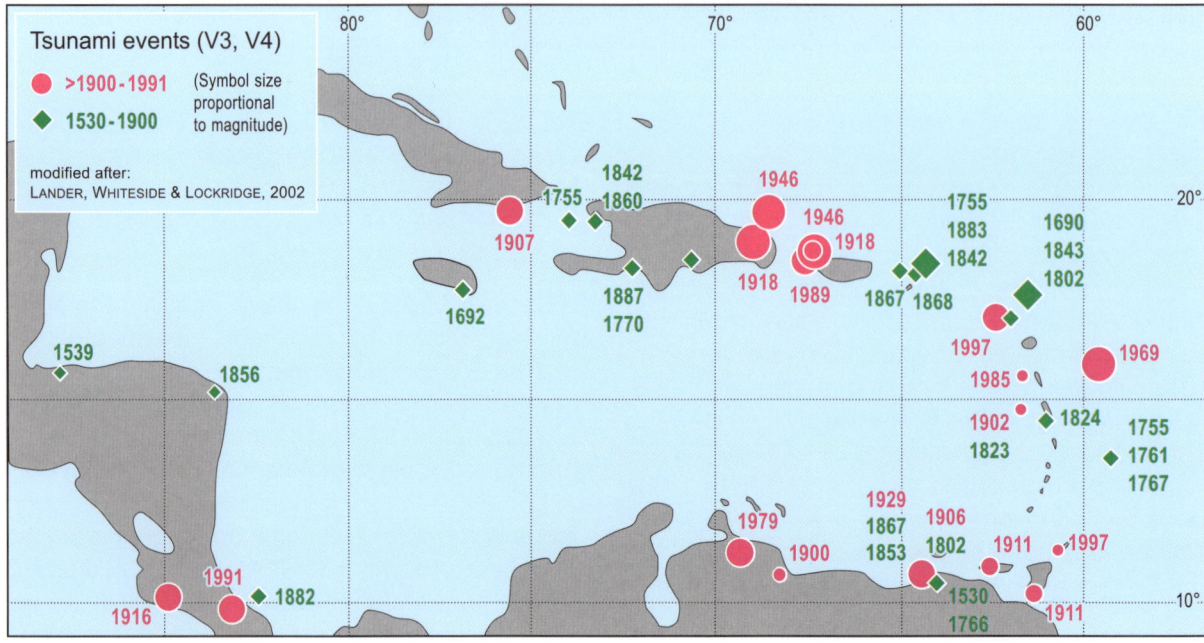

Fig. 13: Tsunami in the Caribbean since 1530 AD (modified after LANDER, WHITESIDE & LOCKRIDGE, 2002). The map shows only tsunami events, which are considered probable (V3) or reliable (V4) by the authors.

tack (see also BUSH, 1991; SCOFFIN, 1993; COCH, 1994; HUBBARD et al., 1991; and others) extreme rainfall (reaching up to more than 1,000 mm within 24 hours, in particular on higher islands) is the most dangerous threat of hurricanes.

1.5 Tsunami in the Intra-Americas Seas

Tsunami have occurred in the Intra Americas Seas in younger times (e.g. 1867 near the Virgin Islands, or 1918 near Puerto Rico), and their history has been compiled first by LANDER & WHITESIDE (1997) and LANDER et al. (2002). Their catalogues show dozens of tsunami since the beginning of the 16th century, some of them with destructive impact (Fig. 13). The causes for the high tsunami risk are tectonic movements along the Caribbean plate boundary as well as volcanic eruptions, caldera collapses or flank instability with slides along volcanic edifices of the Antillean island arc (BOUDON et al., 1999; CARRACEDO, 1999; DAY et al., 1999; DEPLUS et al., 2001; HEINRICH et al., 1999; KELLETAT, 2003; KRASTEL et al., 2001; ROBSON, 1964; ROOBOL et al., 1983; SMITH & SHEPHERD, 1993; WARD & ASPHAUG, 2000; WHELAN & KELLETAT, 2003b, or ZAHIBO & PELINOVSKY, 2001). Submarine slides may have occurred as well (DEPLUS et al., 2001), and the 1755 AD Lisbon event has sent a teletsunami to the Caribbean outer islands. No meteorite impact, as a tsunamigenic agent has been proved for this region, but this cannot be excluded.

All in all there were some theories on tsunami risks in this area, yet the most disastrous tsunami impacts could only be identified by field research on coastal deposits and forms (JONES & HUNTER, 1992; SCHEFFERS, 2002a,b; 2003a,b; 2004; SCHEFFERS et al., 2005a; SCHUBERT, 1994; TAGGART et al., 1993). Thereby we now have proofs for disastrous tsunami of Younger Holocene times (4500 BP, 4000 BP, 3000 BP, 2400 BP, 1500 BP and 500 BP, with nearly 200 absolute data so far) from Grand Cayman, Puerto Rico including the Mona passage west of it, northern Venezuela, Aruba, Curaçao, Bonaire, Barbados, Grenada, Guadeloupe, St. Martin, Anguilla and the Bahaman islands of Cat Island, Eleuthera and Long Island. Together with Japan, the Caribbean can now be designated as the best investigated region concerning tsunami field evidence in the world.

2. Geology and Geography of Bonaire

2.1 Pre-Quaternary Basement

The nucleus of Bonaire island is a ridge of submarine volcanic rocks from Cretaceous to Early Tertiary (KLAVER, 1987; BEETS & MAC GILLAVRY, 1977; BEETS et al., 1984; DE BUISONJÉ, 1974; see also Fig. 14), which later has been topped by subaerial basalt layers, displayed in well developed columns in the Washington-Slagbaai National park. The fact that submarine lavas are now exposed above sea level indicates an uplift of the island since Mid- to Late Tertiary times. Some Eocene marls and clays occur near the village of Rincón. Today these basement rocks reach about 150 m asl in the north,

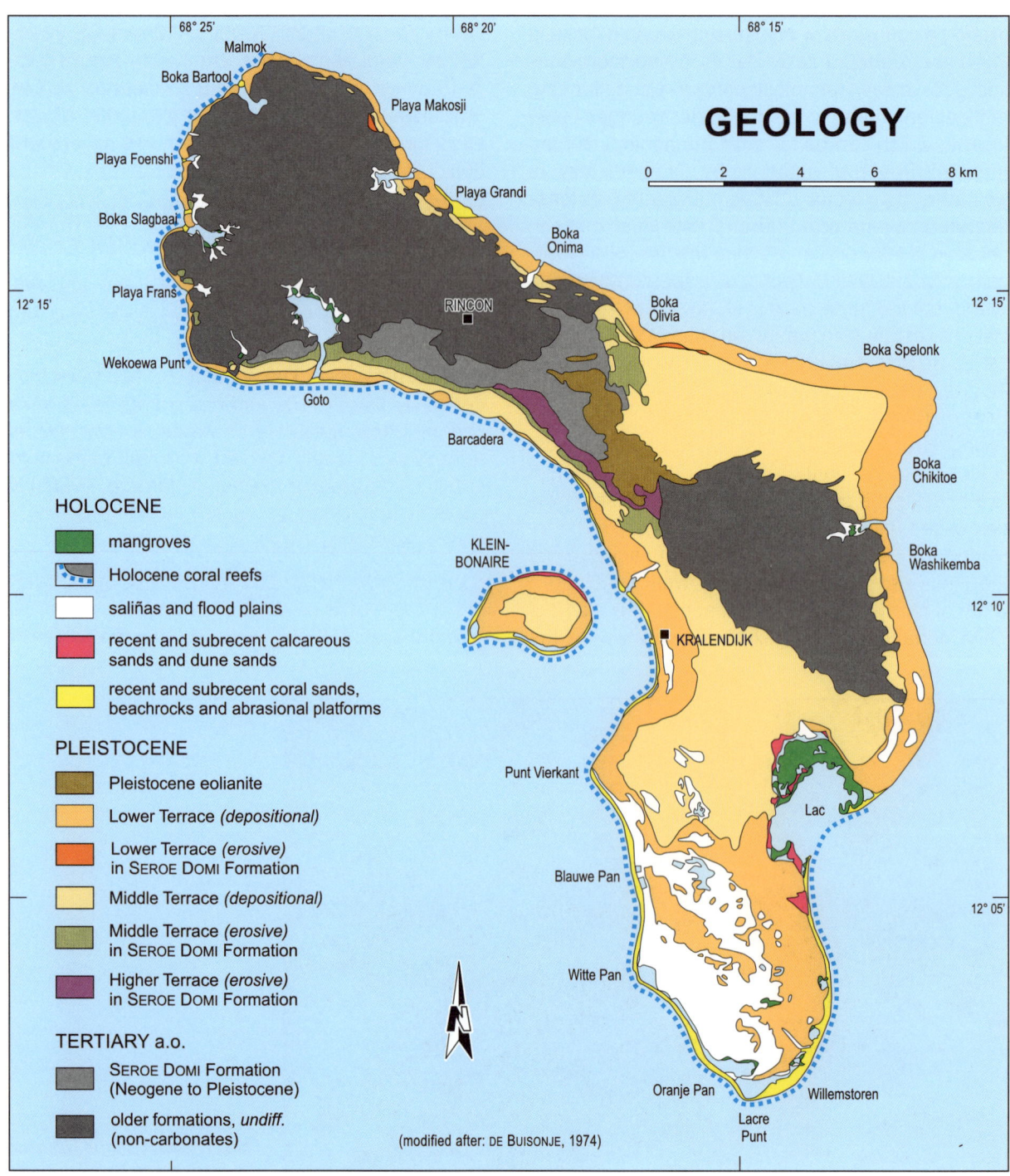

Fig. 14: Geological map of Bonaire, modified after DE BUISONJÉ, 1974.

+20-30 m in the east, and about zero or less in the west and south of the island. The so called Seroe Domi formation of Younger Tertiary age documents, that Bonaire was close to sea level in that period and the first carbonate detritus accumulated on the volcanic rocks. The formation is interpreted as reef detritus.

2.2 Pleistocene Reefs

Since Middle to Young Pleistocene times, Bonaire existed as an island with exposed volcanic rocks in the higher parts, in particular in the northern section. This island was totally surrounded by fringing reefs at least three times during the Younger Pleistocene, which can be seen on the geological map (Fig. 14). Sea level variations by glacio-eustasy as well as a slight tectonic tilting of Bonaire has differentiated several reef bodies. The oldest reef generation is only preserved as a thin discordant cover on Seroe Domi limestone in the center of the island, but the fringing reefs of the last two interglacial phases are well preserved (ALEXANDER, 1961; DE BUISONJÈ, 1974; BANDOIAN & MURRAY, 1974; FOCKE, 1978a; HERWEIJER & FOCKE, 1978; or SCHELLMANN et al., 2004, Fig. 15). Even the youngest reef has been dislocated by tectonics: it is highest in the north with about 15 m asl, only 5 to 6 m asl in the central east parts, and disappears below sea level to the southern cape of Bonaire. During the second last and last interglacial period the reefs developed especially well, reaching a width of many 100 m, and at that time the island of Klein Bonaire began to exist. At many places the coral cover is only 1-2 meters thick, but along the northeastern section it shows layers of reef rock at least 40 m thick.

At Seru Grandi, discordances between reefs of different age are well exposed (Fig. 16), and the reefs themselves show their original terrace like forms, only slightly abraded along the seaward front. It is important to mention that during the last two interglacials reef development along the east and north coasts was not only possible but also flourished. On the last interglacial reef terrace, dated by RADTKE et al. (2002) at 125 ka BP in the highest parts, the species are clearly recognizable: the main species *Acropora palmata* as well as *Siderastrea sidera*, *Montastrea annularis*, *Montastrea cavernosa* and *Diploria* sp. are all present (Fig. 17).

The old reef crests and lagoon areas are distinguishable in the landscape, the first forming a broad and flat ridge, the latter a depression about one meter deep and 100 to 300 m wide (Fig. 18). The lagoon is mostly filled by fine material from the volcanic hills, which enables rain water to stay in these depressions. The original reef has been hardened by diagenesis, and many fragments have changed from aragonite to calcite. A very special feature is a deeply incised and well preserved notch at the base of the higher terrace, evidently cut in by bioerosion during sea level high stand of substage 5e (Fig. 19).

It proves that the recession of the perpendicular old cliff face was close to zero over a time period of more than 100,000 years. Speleothems (cave

Fig. 15:
Well preserved coral (*Diploria labyrinthiformis*) in the Youngest Pleistocene reef.

Fig. 16:
At Seru Grandi in the NE of Bonaire a well preserved discordance separates an older basis reef complex from a younger one, which is at least of isotope stage 9.

Fig. 17:
Well preserved coral *(Montastrea annularis)* in Younger Pleistocene coral reef terrace.

Chapter 2: Geology and Geography

Fig. 18: Slightly uplifted Young Pleistocene coral reef terrace at Washikemba with a well preserved lagoonal area (light brown) to landward.

Fig. 19:
Deep incised notch along the east coast near Boka Chikitu (a) and the west coast at Devils Mouth (b) documenting the sea level high stand of isotope stage 5e in an older reef.

features) in old coastal caves along this notch may preserve a detailed climatic history of the last glacial cycle in this area. A closer analysis of the coral community structure in the Younger Pleistocene reefs show that climate and wind patterns, in particular the tradewind circulation and the position of sheltered and exposed shorelines in the Pleistocene was the same as during the Holocene.

2.3 Geomorphology in General

Bonaire is a low island in the south and center, and exhibits steeper and higher hills only in the north, which is now the Washington-Slagbaai National park (Fig. 20). In this section dry valleys have been incised, only during heavy rain showing floods. Valleys have been more important during sea level

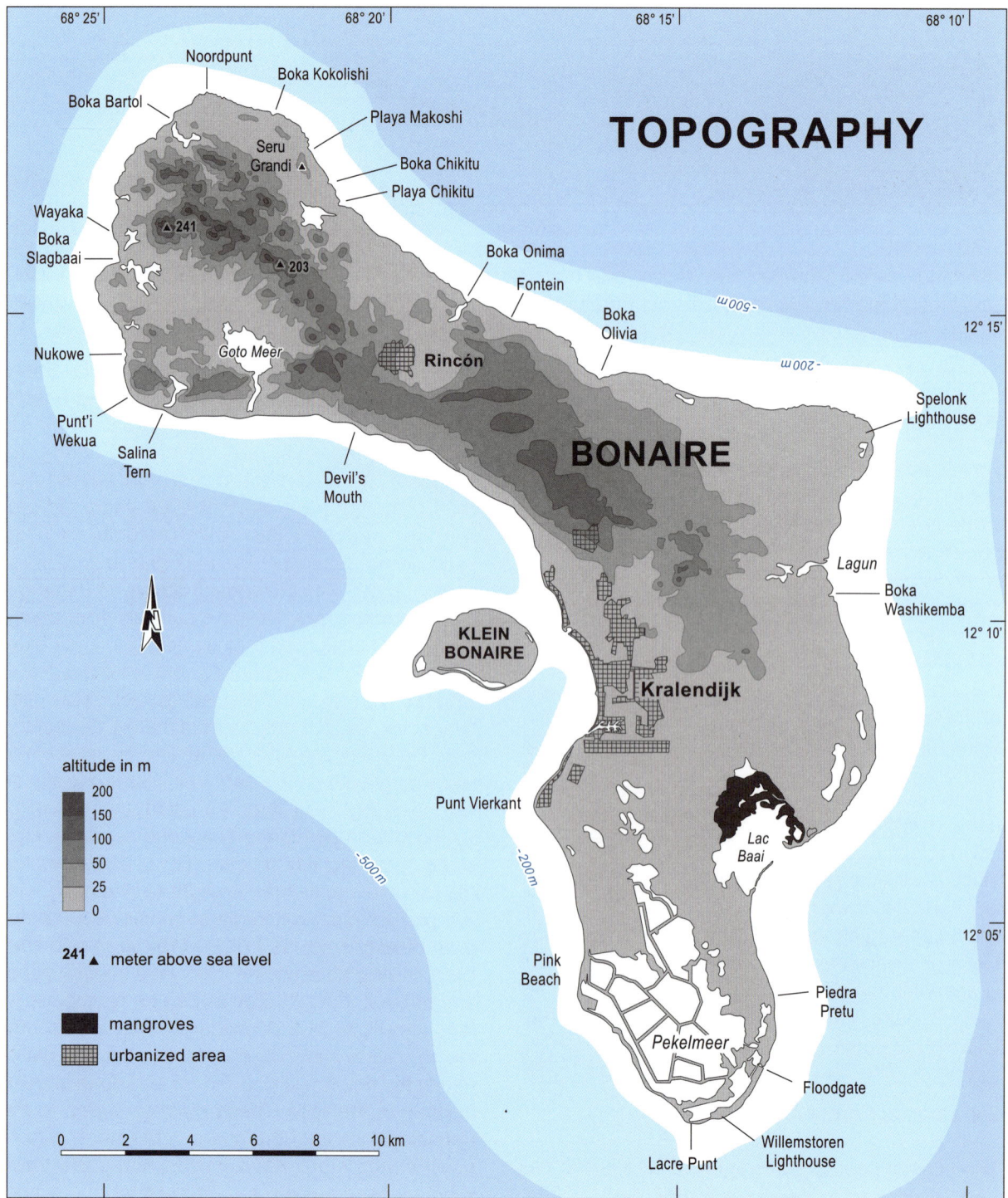

Fig. 20: Topographic map of Bonaire.

Fig. 21:
Artificial salt pans in the south of Bonaire.

lowstands, when Bonaire was larger than today. The most important geomorphological features – beside the dissected hills – are the terraces surrounding the island. These are the former fringing coral reefs. The do not exhibit the differentiation of lagoon and crest such as seen on the reef terrace of the last interglacial, but these terraces show a very slight seaward sloping and are only dissected in few places. They mostly have preserved their steep old cliffs, from which at some places huge boulders are gliding downslope. The southern part of the island is situated very close to sea level and therefore decorated by shallow basins flooded by groundwater and sea water, but in the last decades mostly transformed into artificial salt pans (Fig. 21). They are based by the last interglacial coral reef. Regarding the surface of this reef, a tilting of Bonaire since that time can clearly be detected: The surface of the reef flat built during Oxygen Isotope substage 5e reaches around +15 m in the north, 6-8 m in the northeast, 5 m in the east, drops down to sea level in the south. Along the west coast at Kralendijk the reef flat is around 7-9 m high, further northwest dropping to 3-6 m and rising to the north cape again to more than 12 m. The second last reef terrace is situated about 4-15 m higher but is absent in the southern part of Bonaire (Fig. 14). Klein Bonaire only shows the last interglacial reef exposed to about +3 m in the center.

2.4 Coastal Forms and Processes

The coastal geomorphology of Bonaire island has been described in detail by SCHEFFERS (2002a, 2004), based on observations by DE BUISONJÉ & ZONNEVELD (1960), FOCKE (1977, 1978b,c), KELLETAT (1997), VAN LOENHOUD & VAN DE SANDE (1977) and extended own observations.

Bonaire is mostly a rocky island with cliffs. Shore features are steep or overhanging cliffs with deep basal notches along the seaward front of the second last interglacial reef, developed during the last interglacial high stand (Fig. 19). These notches, although more than 100,000 years old, are very well preserved, often with horizontal roofs several meters leading in the rock and around 1-1.8 m high, partly with old coastal caves. In these caves and notches, Limestone solution from above has formed stalagtites and stalagmites. Their existence, preservation and depth proves the astonishing fact that since the last interglacial, these cliffs they have receded only by centimeters to decimeters. The Holocene cliffs are cut into the front of the last interglacial coral reef, which at most places has been slightly uplifted. Abrasion and in particular bioerosion during the Holocene sea level high stand of the last 6,000 years (RULL, 2000) are the responsible agents for this slight back cutting. The cliff profiles differ markedly depending on the degree of exposure, as FOCKE (1978b,c) and VAN DUYL (1985) have shown. Six wave environments and at least four cliff profiles can be distinguished (see also Figs. 22-24): sheltered west- and south-facing coasts show deep notches with a width of nearly the tide range (0.3-0.5 m) and up to about 5 m deep in the rock. These notches widen to the transitional parts of the coast in the north, where swell brings more wave power. Along the trade wind exposed eastern to northeastern shorelines, where waves generally reach a height of 2.0 to 3.5 m

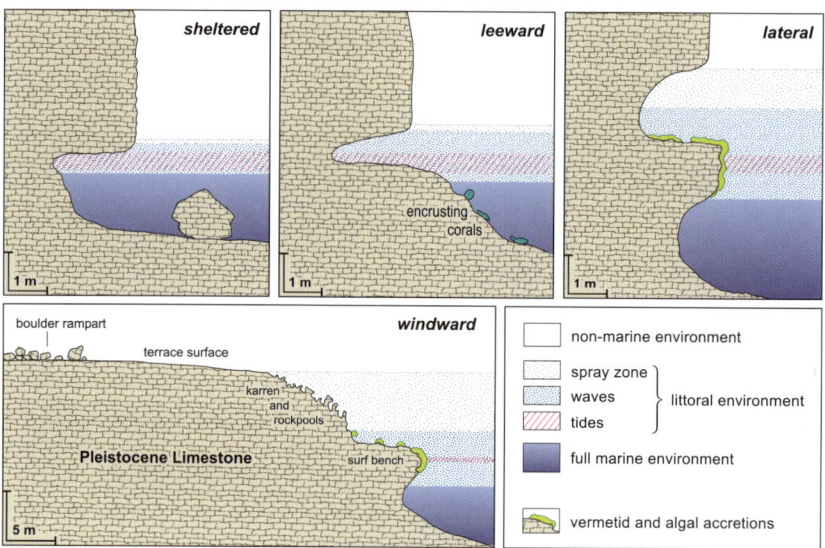

Fig. 22:
Notch types depending on the degree of exposure around Bonaire (modified after FOCKE, 1978c).

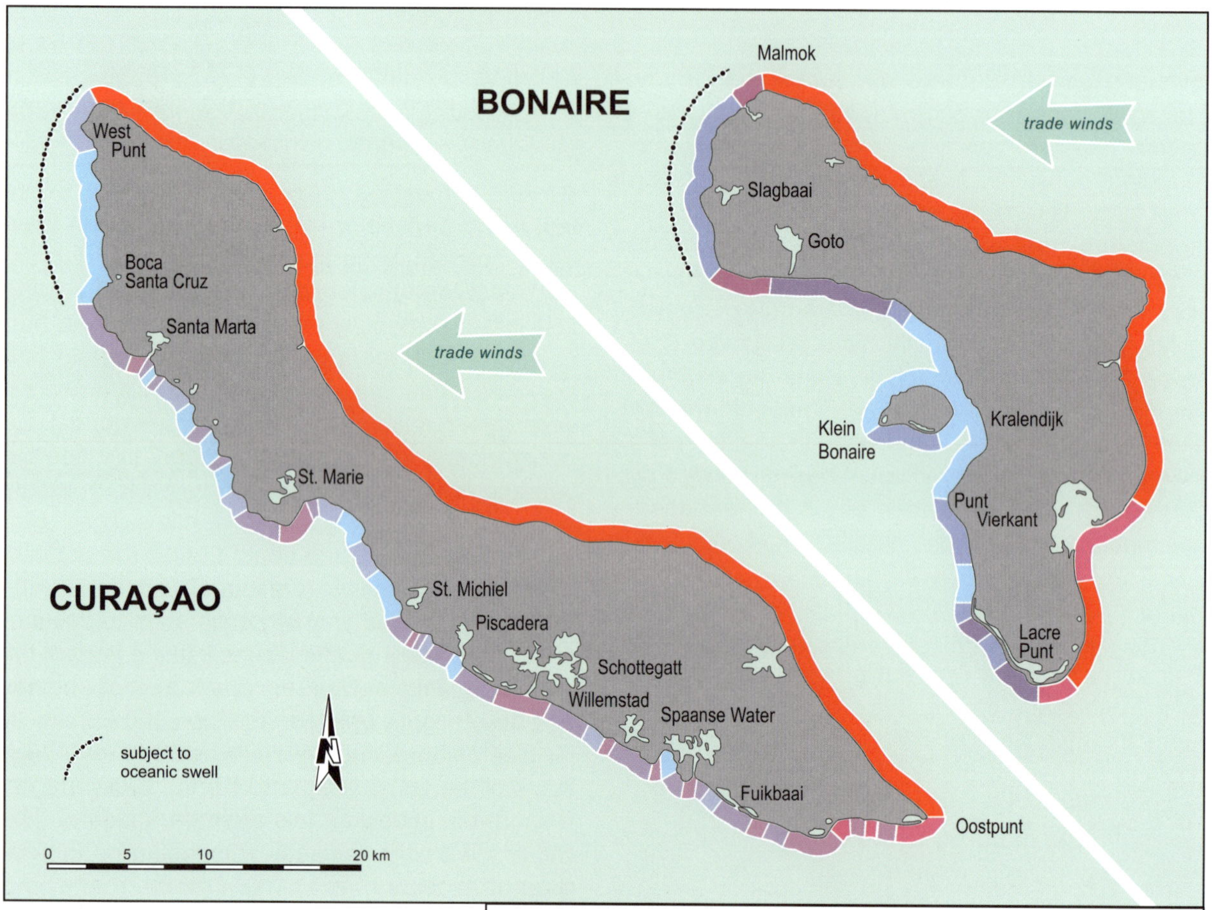

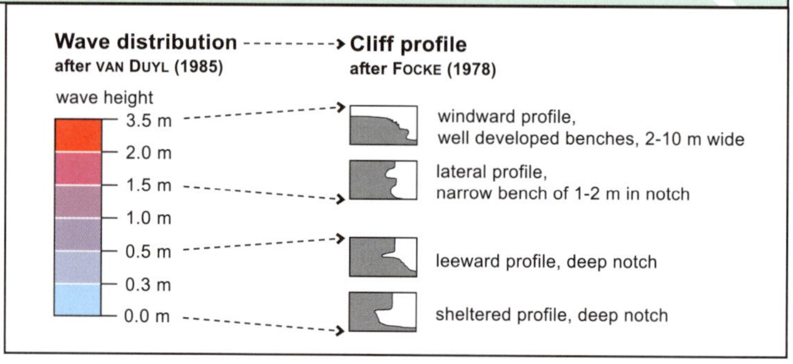

Fig. 23:
Wave distribution and cliff profiles along the Bonaire shorelines (combined from FOCKE, 1978c, and VAN DUYL, 1985).

Chapter 2: Geology and Geography

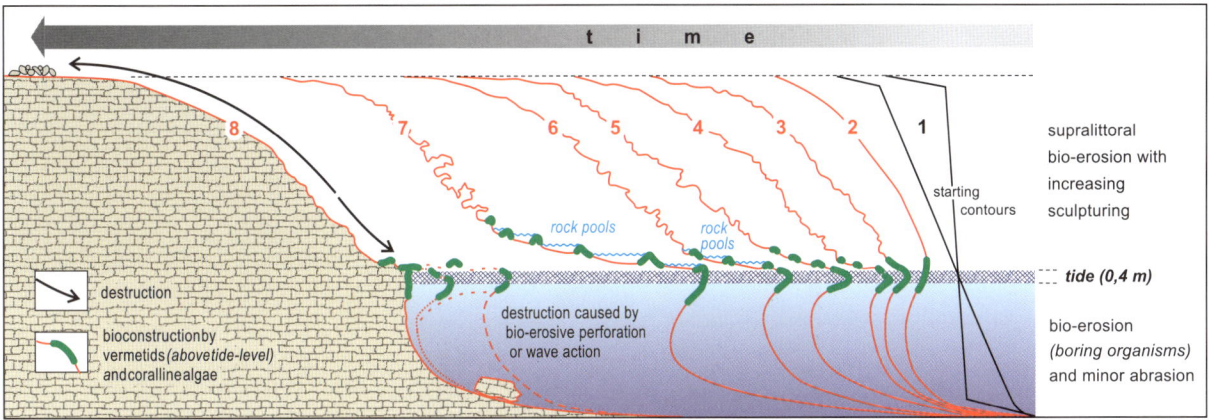

Fig. 24: Development of a cliff profile along the exposed trade wind shorelines of Bonaire (from SCHEFFERS, 2002a).

Fig. 25:
Well developed bench with bioconstructive ledges near Washikemba, east coast.

immediately at the cliff base, notches are not as well developed. More significant is a bench sometimes several meters wide and sloping seaward, separating a broad notch-like basal part along water 5 to 10 m deep and a sharp edged upper shore face, where bioerosion leads to the rock pool belt at the upper supratidal. The bench may be decorated by vermetid rims and ridges (Fig. 25), forming pools filled with sea water by stronger waves and splash. Therefore the bench is partly a destructive and partly a constructive element of the rocky shores of Bonaire. Living organic rocks (the coralline algae *Porolithum pachydermum* and *Lithophyllum congestum* as well as vermetid gastropods like *Spiroglyphus irregularis*) can be found generally up to at least one meter above high water level because of the steady wave impacts from the trade winds along the eastern exposures. Higher than the bench the cliff is almost solely formed by bioerosion. Limpets, littorinids, chitons and other grazers and borers are responsible for this process. A second notch above high water may have developed. The supratidal sensu stricto is the rock pool zone with regular splash and spray, another belt of intensive bioerosion. Pools maybe up to one meter deep and several meters across (Fig. 26), situated in an overall extremely sharp-edged environment. Breaking out of sharp edges or pinnacles left by bioerosion is very seldom and restricted to impacts of extreme hurricane surf or the bombing with particles during storms. Only very occasionally overhanging parts of the cliff break down, if a joint is weakened and opened by karst solution. All in all these mechanical processes are negligible for the rocky shorelines. The percentage of mechanical breakage is even greater along the sheltered coasts, because here bioerosion is not active higher than the notch level because of lack of sea water splash and spray. The intensity of biogenous coastal forming differs extremely between

Fig. 26:
The supratidal rock pool zone at exposed coasts of Bonaire.

the most sheltered sites, where it is restricted to the notch, i.e. a vertical span of decimeters to 2 meters, and the most exposed sites, where it covers a vertical range of about 15 m. Nevertheless the notches with their strictly horizontal extension and rather deep incision are evidence of a long, relative stabile, sea level during the last several thousand years.

Sandy beaches are rare along Bonaire's coastlines. Only at the entrance of Lac Baai (see below) in the southeast, at Lagoen in the east, at Playa Chikitu in the northeast and in the small Boka Kokolishi, sandy beaches exist. All other beaches are built by coral rubble such as seen in small embayments of the west coast, at the rubble bars closing larger coastal embayments (Boka Bartol, Boka Slagbaai, Goto Meer and others), south of Kralendijk to Sorobon, or around Klein Bonaire. These rubble and boulder beaches mostly show ridge like features built by hurricanes or tsunami (see chapters 3.3 and 4.2) and often exhibit beach rocks in different stages of destruction.

The beaches of Bonaire along the southwest and south coast consist of coral rubble, mostly rounded by the surf. Fresh broken-off coral branches are rare, although a fringing reef several decameters wide accompanies these coastlines. Sand produced by bio-erosion on the reef or by abrasion on the beach mostly has been removed from the slope of the reef downwards to deeper water by higher waves. These coastlines are more subject to change by exceptional waves during hurricanes and tsunami, and some of them only show impacts of these events.

Two different types of embayments have to be mentioned for Bonaire's coasts:

The larger ones have a small entrance widening to inland into several shallow arms, showing a former pattern of rivers or creeks now flooded by the postglacial sea level (Salina Term, Goto Meer, and others, see Fig. 27). They belong to the ria forms.

The smaller bokas (embayments) are straight ones, flanked by perpendicular cliffs partly with huge rocks breaking down. These bokas (Boka Washikemba, Boka Kokolishi, Boka Chikitu, and many others), are cut for a length of about 80-120 m into the last interglacial reef terrace (Fig. 28). They mostly end in a small beach interrupted by boulders. Only Boka Kokolishi has been dammed by semicircular vermetid rims bowing against the surf (Fig. 29).

The genesis of these Bokas is a matter of debate (compare SCHEFFERS, 2004): old creeks with fresh water interrupting the built up of an interglacial fringing reefs, cutting off this reef by a younger incision, opening by surf or tsunami waves, break down of coastal caves, fresh water springs to prevent coral growth, are the most probable arguments for their existence.

An exception from this boka type is Lac Baai (dutch for Bay) in the southeast (Fig. 30). This is a rounded shallow bay about 7.5 km² wide with a maximum depth of 4.5 m but mostly shallower at the place of downwarping of the last interglacial. This bay shows a bad developed reef due to the amount of sediment present. Protected by spits at either sides of the entrance continuing into coral reefs and a subtidal bar. Mangroves (mostly *Rhizophora mangle*, partly *Avicennia marina*) are widespread and form a belt up to several 100 m wide

Chapter 2: Geology and Geography

Fig. 27:
Oblique aerial photograph of Salina Term, northwest coast of Bonaire.

Fig. 28:
Oblique aerial photograph of Boka Kokolishi, NE Bonaire.

Fig. 29:
Boka Kokolishi with algal rims.

Fig. 30:
Northern section of Lac Baai with mangroves and coral reefs.

Fig. 31:
Oblique aerial photograph of the narrow fringing reefs of Klein Bonaire.

with shrubs and trees not higher than 6-7 m. This mangrove area shows strong destruction with dead older trees. The reason is not clear: either hurricanes have destroyed these mangroves, or the water regulation scheme of the nearby salines is responsible.

Coastal forming by rare events like hurricanes and tsunami during Holocene times will be discussed in chapters 3 and 4.

2.5 Holocene Coral Reefs

The living coral reefs of Bonaire have been studied in particular by BAK (1975, 1977) and FOCKE (1978d) and mapped at a scale of 1:4,000 in an atlas by VAN DUYL (1985). Ecological aspects, such as the trophic structure have been studied by SCHEFFERS (2005). The reefs fringe with only very few interruptions along the west and south facing shorelines of Bonaire, forming a slightly sloping terrace of 20 to 250 m in width to a drop off at about -8 to -15 m. A steep slope runs down to another terrace at -25 to -50 m, sloping much more to a final drop off into very deep water. Hermatypic coral growth ends between 80 and 100 m of depth. The drop off is decorated by spurs and grooves. These features are also visible in the upper reef terrace, all be it in smaller dimensions (Fig. 31). At the leeward coasts flourishing Holocene reefs, more than 10 m high, have developed, whereas the windward side only shows some patchy and scattered coral communities. In contrast, dense stands of *Sargassum polyceratum* are growing here. During the Holocene, several phases of flourishing fringing reef growth are preserved, and the Younger Pleistocene reefs document a very suitable habitat for reef growth. The question arose why the fringing reefs are missing here nowadays. The idea of VAN DUYL (1985) was, that a slight uplift had narrowed the space for reef growth along the trade wind exposed sides, and that stronger wave impact is the most limiting factor. Along many Pacific islands with steep slopes and strong trade winds, however, we can find flourishing reefs, actively growing laterally against the area of the strongest surf. As will be shown later (Chapter 4.5), during former Holocene times and at least till about 500 BP much more reef growth was present on the shallow terrace along the exposed coasts of Bonaire, which can be proved by the dislocation of a huge amount of coral from that area onshore during extreme events. The riddle of the vanished reefs is addressed in a study by SCHEFFERS et al. (2005).

2.6 Soil and Vegetation

The soil cover of of Bonaire is only of interest in so far as it influences coastal forming, and in particular reef growth. This is the case in run-off scenarios during heavy rain fall, when soil particles may cover the living coral tissue leading to smothering.

As the largest area of soil on Bonaire is represented by brown types with several decimeters of depth and are derived from the diabase and basalt areas of the higher hills, silicate particles represent the largest amount in the run off. The larger ravines and creeks end in bays with still standing water, which are normally closed from the open sea by a barrier of coral rubble like Goto Meer or several natural salinas. Here the fine sediments settle in these bays and mostly do not reach the reef areas. Along the windward side, old lagoon features on the uplifted reefs capture run off particles, and only very few of this suspended matter will reach the ocean. Since nearly no corals are present along this side of the island, there is no damaging effect on reef growth.

Vegetation is another important factor to diminish run off and captures ground water. Several hundred years of colonization and plantage-economy has cleared most of the landscape from a formerly dense vegetation, and only some smaller areas are covered with thorny shrubs with. This vegetation is, unfortunately, not dense enough to hold soil particle on steeper slopes. Extensive grazing by sheep, goat and a rising number of wild donkeys are negative factors for soil preservation as well. These animals even prevent a regeneration of natural vegetation in Washington-Slagbaai National Park in the north.

Fig. 32: Mining of tsunami coral debris at Washikemba.

2.7 Landuse, Infrastructure and Man Made Environmental Impacts

Bonaire is lacking agricultural activities except of some gardening and extensive grazing of goats, sheep and wild donkeys. The most extended area impacted by economical activities isthe solar salt works with many km² of salt pans and a pier for export to the ships in the southern part of the island. Sea water is added to this system by floodgates along the exposed sites, relying on the trade winds to push it into the pans. In these pans, the sea water circulates until it reaches a specific density for the saturation ponds, in which the water evaporates until only the salt cristals are left. This salt plant has been developed during the sixties of the 20th century. An aquaculture farm has been established near the southern entrance to Lac Baai.

Another extended and landscape consuming activity is mining of coral debris, devastating the coastal landscape for kilometers along the eastern and northern shorelines. Here, on the flat uplifted coral terrace, coral debris (mostly from tsunami, see chapter 4 and Fig. 32, see previous page) is easily accessible by trucks. The debris is crushed into sand, which is used for constructions (Fig. 33). Although Bonaire has only about 15,000 inhabitants, there is a lot of development along the sheltered coastlines in the north and south of Kralendijk, the center of economic activities including harbor facilities and a cruise ship terminal. In the north along the west coast near Goto Meer, a petroleum depot with pipelines to ships in deeper water has been established. The trade wind exposed coastlines in the east and north, however, have not been developed because of strong wind and steady salt spray. Except of two restaurants and less developed small sandy beaches in southern Lac Baai no infrastructure is present.

Compared with the neighbouring islands of Curaçao and Aruba, the coral reefs of Bonaire's leeward side are better developed and belong to the world's best diving sites. These reefs are protected by a Marine Park (STINAPA). Inexperienced divers, however, are a steady threat to the living coral reefs here.

Fig. 33: Coastal infrastructure around Kralendijk, the main town of Bonaire.

3. Storms and Hurricane History of Bonaire

3.1 The Historical Perspective: Storm and Hurricane Tracks of the Last 150 Years

The Leeward Netherlands Antilles, consisting of Bonaire, Curaçao and Aruba, situated at 12 degrees north lay outside the hurricane belt (Figs. 12 and 34-36). The amount of tropical rain brought by these storms does not harm the islands much, partly because of the limited amount and shorter duration as compared closer to the eyes of hurricanes, partly because of the limestone rocks soaking up a large amount of groundwater. For this region a list of storms and hurricanes for the last 150 years does exist, as well as written information from different

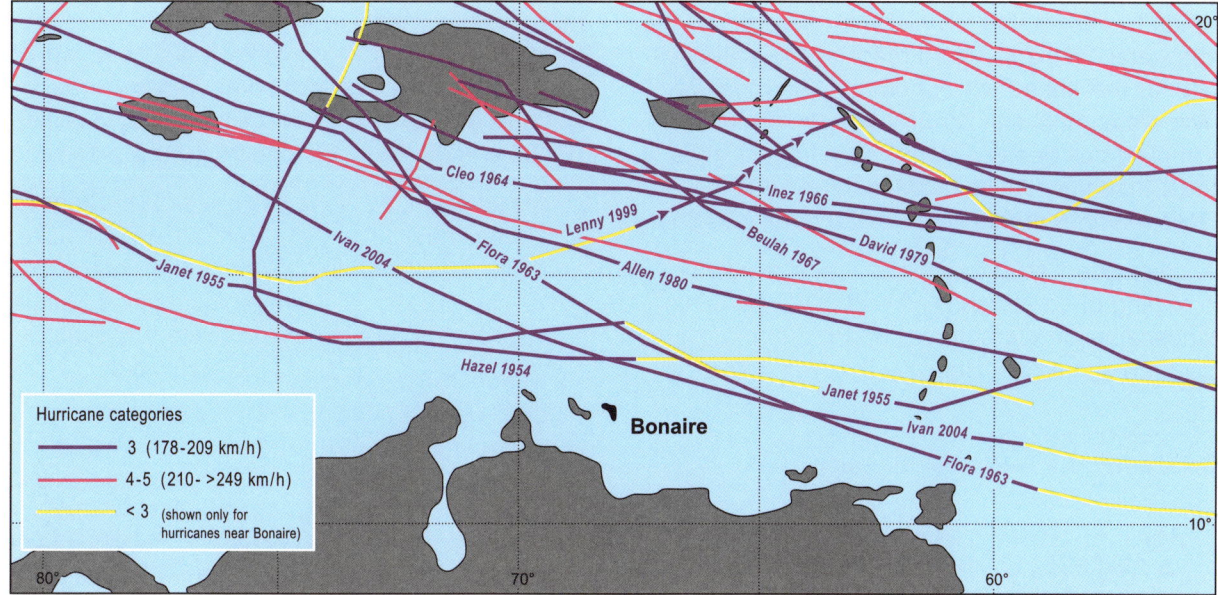

Fig. 34:
Tracks of hurricanes of category 3-5 for the time period 1851-2004. Hurricanes close to Bonaire display names (from NOAA, 2004).

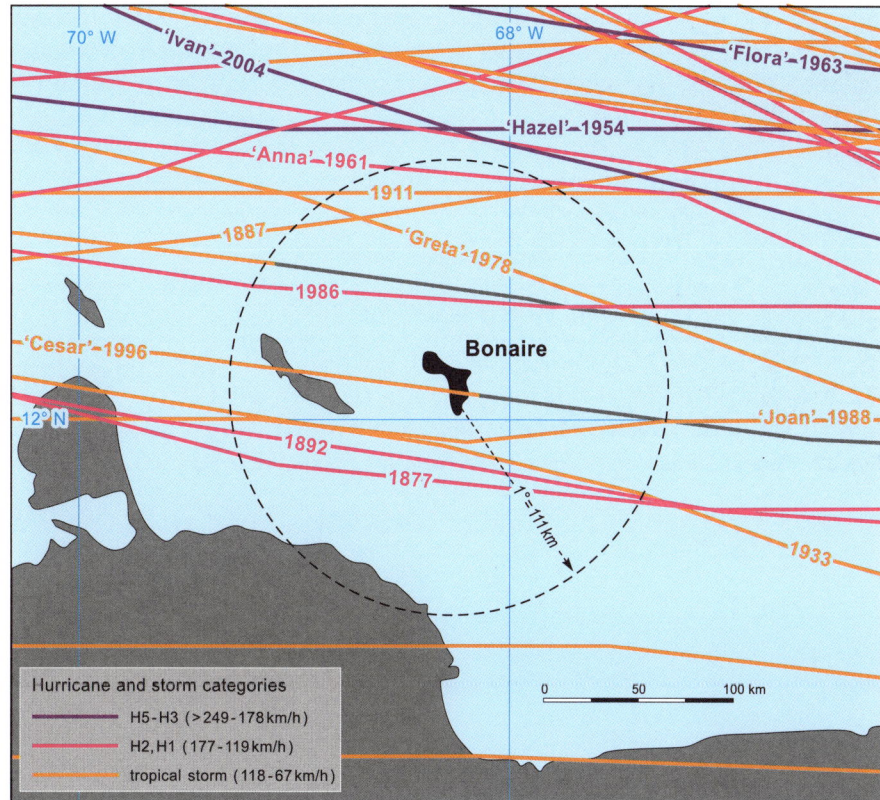

Fig. 35:
Ten tropical storms and hurricanes up to category 2 came into a distance of 60 nautical miles (= 111 km) of Bonaire from 1851 to 1998. The strong hurricane *IVAN* passed just north of this radius (STORMCARIB, 2005; NOAA, 2004).

Table 1: The SAFFIR-SIMPSON Hurricane Scale.
Wind speed is the determining factor in the SAFFIR-SIMPSON scale. It may be combined with storm surge, but the values of this effect are depending strongly on coastal configuration, atmospheric pressure, speed of the hurricane system as a whole, and foreshore topography. Wind speed is measured during sustained winds (meaning an average speed of at least one minute duration). Gusts may be much stronger.

Category	Winds [km/h]	Storm Surge [m]
1	118-153	normally 1.2-1.5
2	154-177	normally 1.8-2.4
3	178-209	normally 3-4
4	210-249	normally 4.2-6
5	> 249	greater than 6

nations sailing in this area since 500 years. From these sources only few information on destructive impacts of hurricanes on Bonaire in historical times can be derived. Fig. 34 shows tracks of 25 storms and hurricanes with category 3 to 5 between 1871 and 1999, all situated well north of Bonaire. Ten tropical storms and hurricanes up to category 2 (Fig. 35), however, passed the island within a distance of only 60 nm (nautical miles), which is equivalent to about 111 km in the years 1877 to 1996, and Hurricane *LENNY* with category 4 in 1999 as well as hurricane *IVAN* with category 4 to 5 in 2004 show that Bonaire, even though it lays outside of the hurricane belt, is threatened by hurricanes.

3.2. Hurricane *IVAN* of September, 2004

The year 2004 was a very special one in the hurricane history of the Caribbean, Florida and the south coast of the United States: Five major hurricanes occurred, and three of them have had the category 4, two partly category 5 with sustained winds of 250 km/h and more. They all came from the open Atlantic. The last in this line was hurricane *IVAN* mid of September, 2004. It was a category 4 hurricane all along from outside the Antillean island arc to the south coast of Louisiana and Alabama (USA), its winds affected 85% of all structures on Grenada island, and the rain caused several landslides. Although the eye of hurricane *IVAN* passed only 140 km north of Bonaire (Fig. 36), rainfall was rather limited here and restricted to about 9 hours, and the winds did not surpass 60 km/h. The waves, however, arriving as a wide swell from the east midday of September 15th, rose to at least 12 m along the cliff coast of Boka Onima (Fig. 37) during the late afternoon of that day and lasted until the morning of September, 16th. These are several thousands of extreme high waves to hit the easterly coasts of Bonaire island. Water fountains more than 30 m high rose along the cliffs, and the sheet of water rushing up the reef terrace at +5 to +6 m could reach more than 1.5 m in thickness (Fig. 38). The inundation therefore was several 100 m on the old reef flat, which includes a lagoonal depression, and dead fish could be found high above sea level and far inland (Fig. 39).

As eyewitnesses of these exceptional hurricane waves (compare also SCHEFFERS et al., 2005b) we inspected the coastlines of Bonaire to look for change in the landscape, either depositional or destructive. As the waves came from the east, which is the exposed trade wind side with cliff coasts and nearly without any infrastructure, the

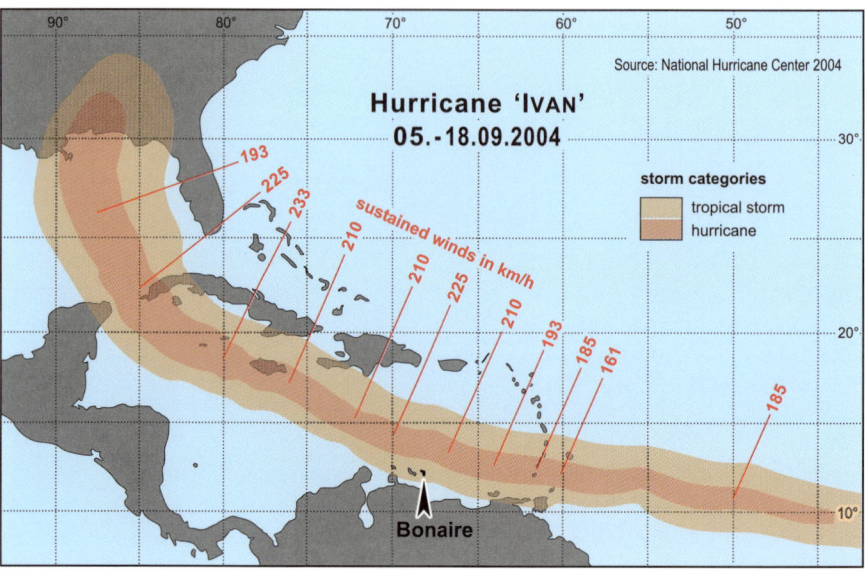

Fig. 36: Track of hurricane *IVAN*, September, 2004.

Chapter 3: Storms and Hurricane History

Fig. 37:
Wave crest of hurricane *IVAN* swell along the east coast of Bonaire on September, 9th, 2004, with a height of about 10 m asl.

Fig. 38:
Backwash of *IVAN* waves at a coral reef terrace 5-6 m high.

damage was limited, and natural phenomena were the most distinctive imprints. The sheltered leeward coastline at Willemstoren Lighthouse at the southern tip of the island to the northwest was left untouched, all be it with some exceptions: The spit newly created five years ago by hurricane LENNY north of Slagbaai was washed over, broadened and eroded to about 1.3 m, and its end was curved against the mainland for several meters (Fig. 40). Nearby at the beach of Wayaca III the old coral rubble ridge created by LENNY (1999) had been pushed up to a new and steeper ridge up to 1.5 m high. At Playa Frans small platy boulders were leaning against trees or are distributed in the bush vegetation. Some fresh coral debris had been deposited on the cliff tops at +6 m north of Slagbaai, but in a lesser amount than from hurricane LENNY (which came from the opposite direction). The ridge from hurricane LENNY (1999) in Nukowe Bay was pushed up, the crest was sharpened and shifted inland. Fragments

COASTAL RESPONSE TO EXTREME WAVE EVENTS

Fig. 39:
Dead fish could be found more than 100 m inland at 6 m asl after hurricane *IVAN* waves.

Fig. 40:
The spit newly created in 1999 by hurricane *LENNY* north of Slagbaai has been flattened, broadened and curved to inland by swell from hurricane *IVAN*.

Fig. 41:
Sand from old bimodal tsunami deposits have been washed out by *IVAN* waves and redistributed on the higher reef flat.

found had diameters of 30-40 cm and in the case of *Meandrina meandrites* more than 50 cm.

Four effects of hurricane *IVAN* waves are more significant than these singularities:

1. Mining of the extended coarse tsunami debris along the exposed shorelines of Bonaire has left finer grains sands and broken shells. The *IVAN* waves redistributed these sands in a wide amount, forming fields of ripple marks (Fig. 41) and covering lowly vegetation. This streaming process on the nearly plain surface was only able to move particles of about 1-2 kg. In the area along Seru Grandi in the northwest of Bonaire, where a lot of sand have been deposited by trade winds from the sandy shore of Playa Chikitu (Fig. 42), embedding large old tsunami boulders, fragments have been excavated partly from the sand, now exhibiting a lower part without the dark grey coating of endolithic algae (Fig. 43).

The difference in color along single boulders documents a long time of a relative stable situation in sand cover before hurricane *IVAN*. As direct observations could prove, nearly no fresh sand has been deposited on the old reef surface, because the *IVAN* waves were not able to lift any significant amount of sediment from the foreshore. All what could be seen in the high splash were fragments of *Sargassum* and other soft algae.

2. As SCHEFFERS (2002a,b, 2004) has described and documented, the extended coarse rampart sediments from dated tsunami on the ABC islands are separated from the cliff edge by a belt

Fig. 42:
Along Seru Grandi in NE Bonaire wide sandy flats have been washed over at least till 9 m asl from *IVAN* waves.

Fig. 43:
At some places old sand covers have been washed out into the sea, leaving marks of former height along old tsunami boulders.

Fig. 44: Seaward of the tsunami boulder ramparts a sediment free belt has been created by many strong hurricanes.

Fig. 45: The rock pool belt shows some small destructions but no sediments in the traps after *IVAN* has passed.

Fig. 46:
Along the south coast of Bonaire east of Willemstoren Lighthouse *IVAN* waves have formed a new broad ridge up to nearly 2 m high, mostly using old tsunami rubble.

of sediment free rock (Fig. 44), mostly decorated along the sea side by bioerosive rock pools, which themselves are free of sediments although being excellent traps for all kind of debris. A first idea was, that this is the original deposition pattern, as tsunami have sedimented the coarse debris in a certain distance from the surf, and storms evidently were not able to bring coarse particles on top of the cliffs 4-6 m high in this exposition. By observing the waves of hurricane *IVAN* we learned more about the phenomenon of the sediment free belt: it is reached by the strongest storm waves, which run against the tsunami debris ramparts or ridges. The backwash from these waves takes all medium to small particles out into the sea clearing the frontal reef terrace by this process. At some places such as Boka Onima (and at the northeasternmost tip of Curaçao at +8-9 m) *IVAN* waves were able to shift the tsunami ridge several meters to inland and steepen its seaward front. The backwash as a water film of about 0.5 m was strong enough to remove all fragments trapped in rock pools and left them sediment free. By transporting coarser fragments seaward, sharp edges of the bioerosive sculptures between the rock pools have been broken off (Fig. 45), and small edges with the light colour of the coral limestone appear on the overall dark surface. The overall effect of the *IVAN* waves along the cliffed eastern and northern shores of Bonaire has been erosive, bringing a certain amount of sand and debris back into the sea. Along the southern tip of Bonaire east of the Willemstoren lighthouse *IVAN* waves have abraded the broad tsunami rampart (which was partly more than 2.5 m thick) for some meters. In places where the rampart was not higher than 1.8 m above sea level, *IVAN* waves could cover the seaward parts with fresh debris from the beach zone, and in lower parts the old ridges have been reworked (Figs. 46-48).

3. At the most exposed sites, the surf of hurricane *IVAN* was able to destroy vertical or overhanging parts of the limestone cliff: sections more than 20 m long, 1.5 m wide and 2-3 m high have been broken down, hitting the bench and disappearing in deeper water (Fig. 49). Some of these destructions have certainly moved several 100 tons at the same time. But there are only very few places where waves have transported fragments from the cliff edge to inland. One example can be seen south of the Spelonk Lighthouse (Fig. 50), and another example in Fig. 51shows a boulder from the sublittoral (proven by the fresh borings of bivalves or sea urchins). The weight of this boulder is about 6 tons, deposited 30 m from the sea at a height of +5 m, i.e. moved against gravity for at least 11 m. Large platy boulders are removed from the old coral reef terrace (Fig. 52), and a mushroom rock of 22 tons in Boka Chikitu (Fig. 53) has been broken off its base.

4. Regarding the ongoing debate, whether large boulders onshore have been deposited by storm waves or are an undisputable proof for a tsunami, hurricane *IVAN* has given some good lectures. Although *IVAN* produced some of the largest waves (at shore line level at least 12 m high, see Fig. 37) ever reported, the capability of these

Fig. 47:
Very good imbrication of the old tsunami ridge is still preserved even in areas of strong *IVAN* waves.

COASTAL RESPONSE TO EXTREME WAVE EVENTS

Fig. 48:
Storm waves may alter the morphology of an older tsunami ridge depending on its height above sea level.

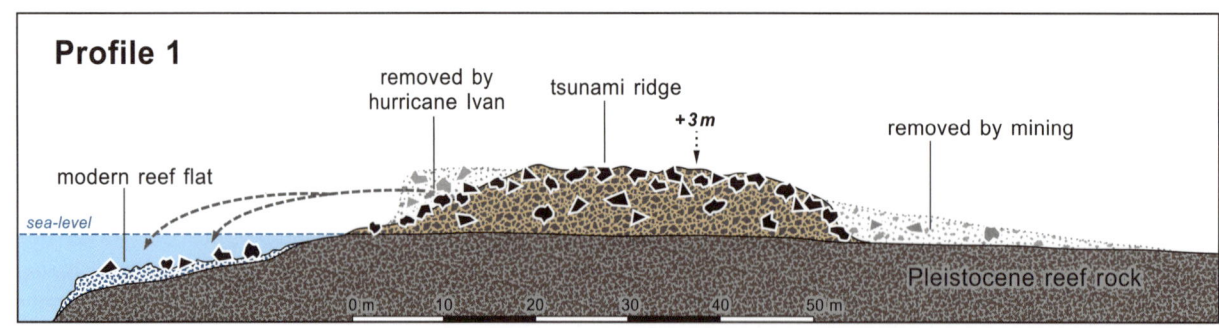

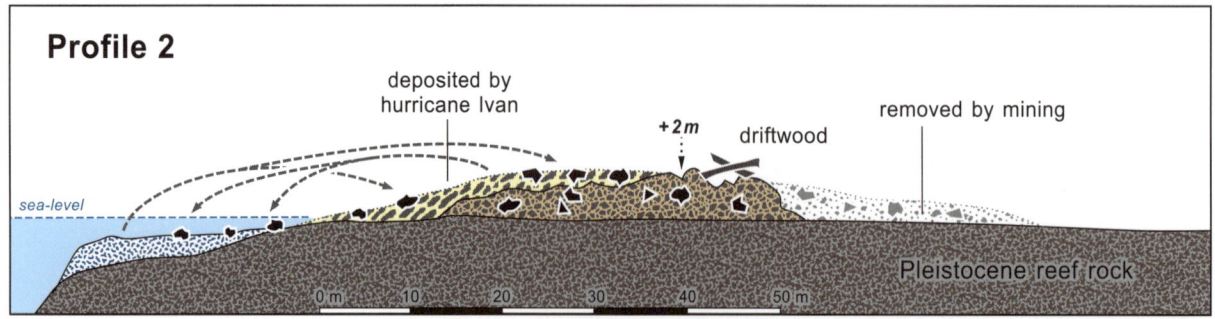

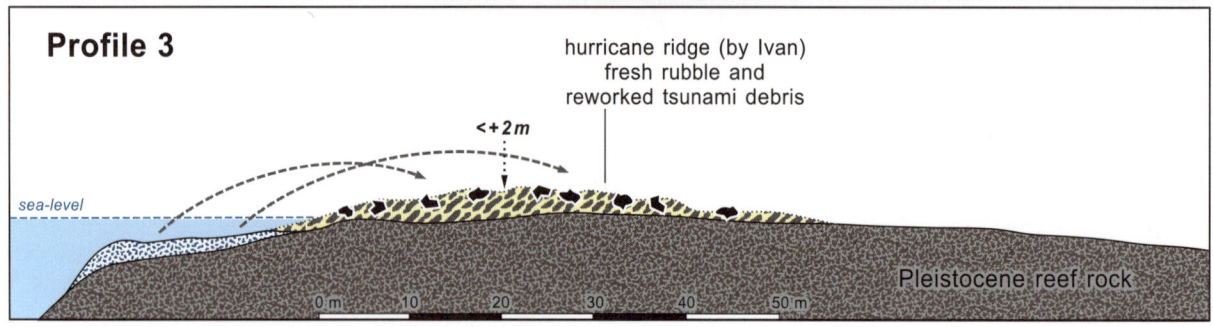

Profile 1
- modern reef flat
- removed by hurricane Ivan
- tsunami ridge +3 m
- removed by mining
- Pleistocene reef rock

Profile 2
- deposited by hurricane Ivan
- +2 m
- driftwood
- removed by mining
- Pleistocene reef rock

Profile 3
- hurricane ridge (by Ivan) fresh rubble and reworked tsunami debris
- < +2 m
- Pleistocene reef rock

Willemstoren Lighthouse

Chapter 3: Storms and Hurricane History

Fig. 49:
The cliffs along the east coast of Bonaire show break off in parts, and some destruction can be seen in the rock pool belt from *IVAN* waves.

Fig. 50:
One of the larger boulders, broken off the cliff by *IVAN* waves. Weight is around 6 tons.

Fig. 51:
The largest boulder from the sublittoral, weighing 6 tons, now at +4-5 m asl and moved at least 12 m against gravity by *IVAN* waves.

waves to move large boulders was very limited. *IVAN* waves have moved far less than 1‰ compared to one of the Holocene tsunami of Bonaire. New boulders added onshore by *IVAN* did not exceed 6 tons, and older tsunami boulders have been moved up to a weight of 50 tons, but only for decimeters or a maximum of 1-3 meters (Figs. 54-59 and Table 2), whereas the tsunami boulders may reach weights of more than 100 or 200 tons and about 400 tons in the case of Boca Chikitu at the east coast of Washington National Park.

This is roughly in agreement with the calculations of NOTT (1997, 2003a,b) for wave heights required to overturn boulders at sea level: boulders of more than 20 tons require storm wave heights which do not occur along the world's shorelines. An instructive example of *IVAN* wave power can be given by the mushroom rock of Boka Chikitu (Fig. 53): It has a weight of more 22 tons, and its bioerosive notch is up to 1 m deep, pointing to an age of this form of many hundred up to one thousand years. This rock has been broken from the sea floor, tilted and moved several meters land inwards but still close to the shoreline. Regarding the depth of the notch the rock has withstand all waves for many 100 years, but hurricane *IVAN* waves were able to destroy it, pointing to their extraordinary power.

To conclude: there is a significant difference in nature between the wave power of a category 5 hurricane and a tsunami, both regarding the destructive forces as well as the debris deposits or the weight of dislocated boulders. It is in the order of 1 to 10 or more for boulder weights, and about 1 to 1,000 for the amount of material moved.

Fig. 52:
A platy fragment of about 20 m² has been broken from the old reef terrace.

Fig. 53:
A mushroom rock in Boka Chikitu with a weight of 22 tons has been broken off and dislocated for several meters.

Chapter 3: Storms and Hurricane History

Table 2: Large boulders moved by waves of hurricane *IVAN*, September, 2004 on Bonaire (in order of their distance from the shoreline).

Distance of present position to shoreline	Altitude [m asl]	Weight [t]	Horizontal dislocation [m]	Vertical dislocation [m]	Kind of movement	Transport figure [1]
6	0	**22**	8	1	broken off eulittoral	176
10	5	**2**	15	12	broken off sublittoral	360
10	5	25	4	1	pushed	
10	5	87	1	1	lifted	
15	2	**5**	7	4	broken off sublittoral	140
30	5	**0.5**	35	11	broken off sublittoral	192.5
30	5	9	10	1	gliding	
35	4	**2**	40	6	broken off sublittoral	480
45	6	3.5	16	1	pushed	
60	5	**3**	70	12	broken off sublittoral	2,520
70	5	**6**	25	12	broken off sublittoral	1,800
100	8	3	64	2	gliding	
110	5	30	2	1	lifted	
110	5	40	1	1	tilted	
110	5	50	10	1	pushed	
120	5	12	3	1	overturned	
120	5	30	3	1	overturned	
125	5	37	2	1	overturned	
140	5	3	6	1	pushed	

Broad type: freshly broken off boulders, Normal type: pre-existing boulders, only moved by *IVAN* waves
[1] Transport figure = weight in tons × distance moved × vertical distance (only for fresh *IVAN* boulders)

Fig. 54:
Old tsunami boulder of more than 20 tons has been tilted at +5 m and more than 100 m distant from the cliff.

COASTAL RESPONSE TO EXTREME WAVE EVENTS

Fig. 55:
This tsunami boulder with a weight of about 40 tons has been uplifted and settled down on some smaller rocks.

Fig. 56:
Delicate new setting of an old tsunami boulder 110 m apart from the sea.

Fig. 57:
These two tsunami boulders have been tilted into an upright position by *IVAN* waves, although weighing 40 to 50 tons.

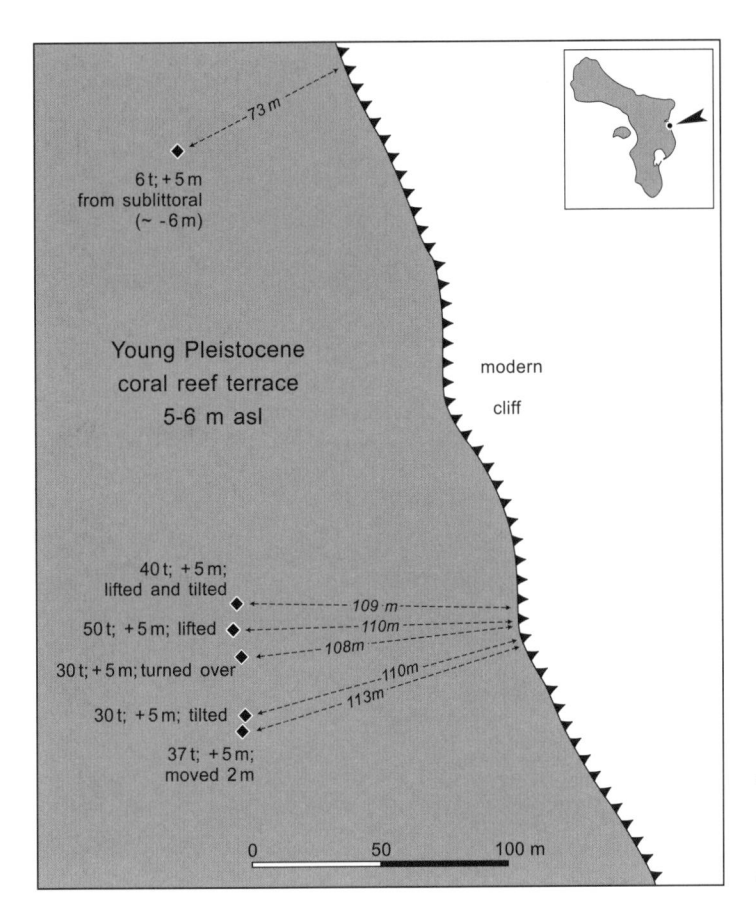

Fig. 58:
This sketch shows the distance of large tsunami boulders from the modern cliff, but moved by hurricane *IVAN* waves along the Washikemba coast east of Bonaire.

3.3 Hurricane *LENNY* of November, 1999

Of all storms and hurricanes known in the 120 years long storm history of Bonaire only one came from the west. It was hurricane *LENNY* in November, 1999, and was the youngest strong storm before hurricane *IVAN*. It passed the island in a distance of about 200 km (Fig. 60). With more than 200 km/h sustained winds (it was categorized 4 according to the SAFFIR-SIMPSON Scale), *LENNY* was able to create a wide field of high waves hitting Bonaire (Aruba and Curaçao as well) from the west and northwest and refracting along the leeward coast to southerly exposures. The small island of Klein Bonaire was also hit along its northern and western shores. Before *LENNY*, approximately 100 years without extreme wave impacts guaranteed uninterrupted coral growth on Bonaire's leeward fringing reefs. With wave heights of 6 or even 8 m *LENNY* produced the largest waves on record for this region, which only have been surpassed by the *IVAN* waves up to 12 m some years later, but at easterly exposures. The *LENNY* waves damaged fringing reefs along the leeward sites to depths of -5 to -10 m (average damage 24.4% of all corals, compare BRIES et al., 2004). Even coral pinnacles and head corals of more than one meter in diameter (which represents ages of more than 100 years) have been overturned at some places. The uppermost parts of the reefs, which suffered from smaller storm wave impacts, withstood the crushing wave of *LENNY* better than the deeper ones, because in water deeper than 5 m no significant wave impacts occurred for a very long time. *LENNY* added some remarkable new coastal features to Bonaire's shores: fresh beach ridges up to 1 m high and up to more than 8 m wide with approximately 20-30% of fresh broken coral and the rest as rounded and bored coral rubble (Figs. 61, 62).

LENNY ridges are characterized by the tongue-like ramps land inward (Fig. 63), where they at several places partly buried mangrove shrubs. On Klein Bonaire, however, the ridges show these features only in a less developed form.

Beside new beach ridges, *LENNY* created a spit nearly 100 m long, 4 m high (about 1.6 m above the water) and more than 10 m wide at the base, grown from a small rocky promontory just north of the Slagbaai buildings (see SCHEFFERS, 2002a, and Fig. 64). The buildings have been partly destroyed by the *LENNY* waves (Fig. 65), and rangers of the Washington-Slagbaai National Park told that the wave height was up to 8 m in this area. *LENNY*

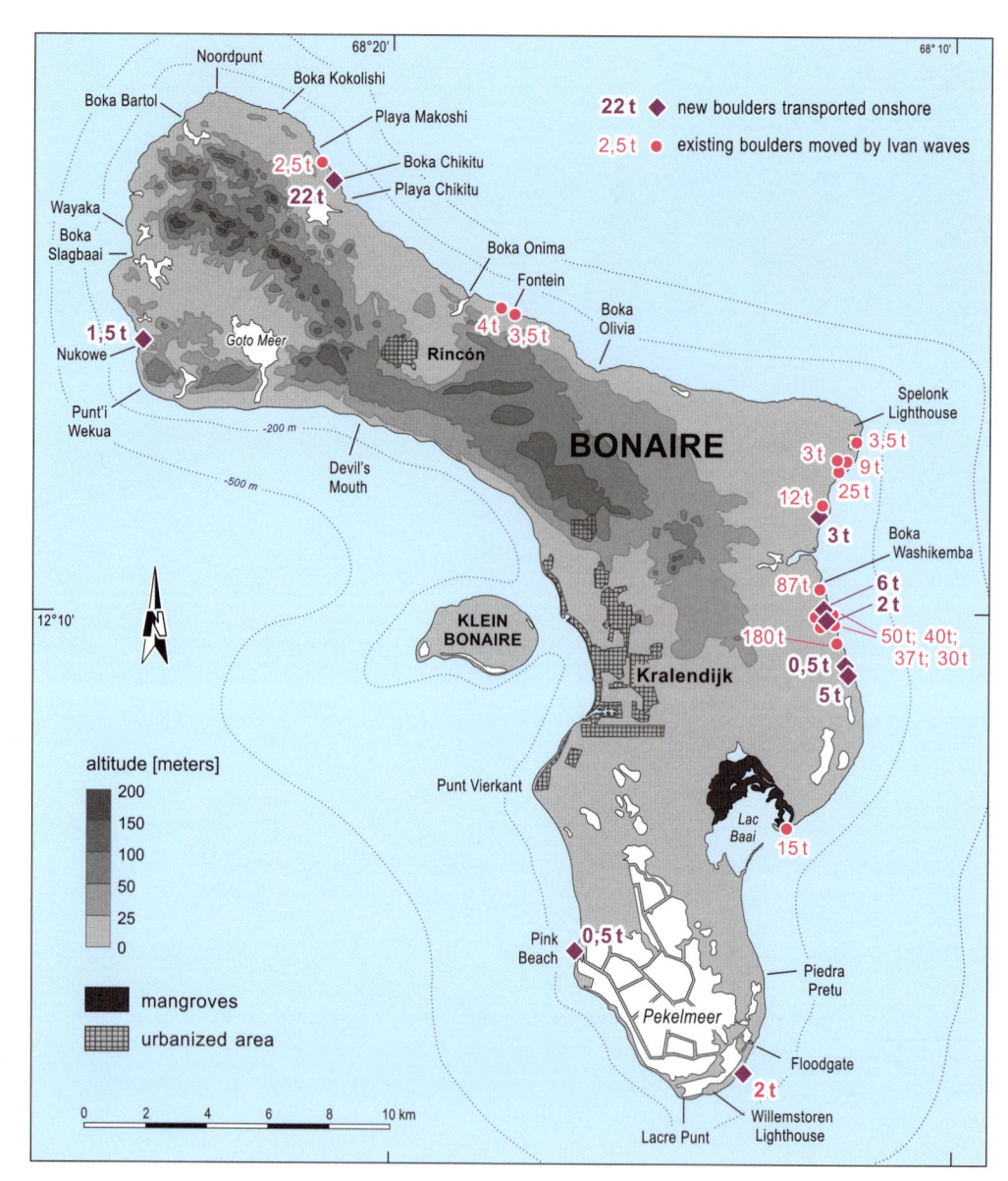

Fig. 59:
Distribution of large boulders around the coastlines of Bonaire, moved by waves from hurricane *IVAN*.

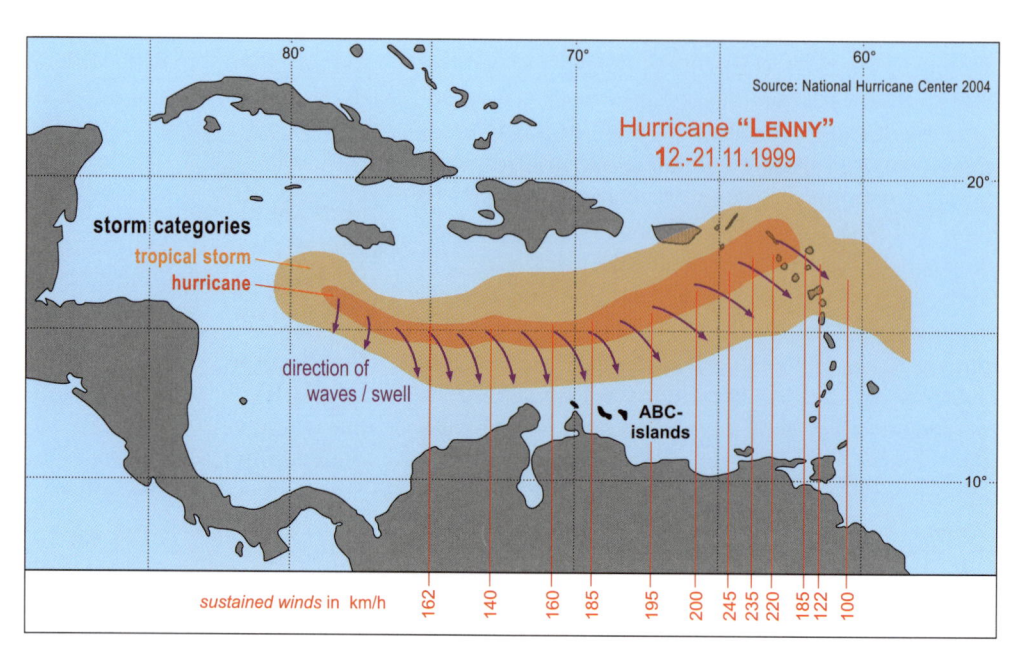

Fig. 60:
The unusual track of hurricane *LENNY* (November, 1999) from the west.

Chapter 3: Storms and Hurricane History

Fig. 61:
Fresh hurricane *LENNY* ridge in Nukowe Bay, NW Bonaire. Darker debris on the left are from a storm event in 1877.

Fig. 62:
Fresh storm ridges from *LENNY* in Nukowe Bay.

Fig. 63:
The *LENNY* ridges show debris tongues to inland, partly burying mangrove bushes.

43

also filled small gaps in the cliff up to +5 m with coral rubble (Fig. 66) and built a new terrace at about +1 m parallel to that of the storm of 1877 (see chapter 3.4 and Fig. 67). The LENNY hurricane waves adapted to smaller variations in the vertical cliff (with deep notch) 5 m high in the way that they deposited semi-circular boulder deposits on top of the last interglacial coral reef terrace (Fig. 68). At single places, boulders up to more than 100 kg have been transported up to nearly +6 m and more than 30 m inland (Fig. 69).

Beside the new spit and beach ridges LENNY waves were able to move some large boulders: one example can be seen on Klein Bonaire, where two boulders, about 3 tons each, have been uplifted from the base of a cliff about 1 m high and deposited just on the cliff top. Another example is a boulder of about 12 tons on a cliff top on the east coast of Curaçao at about +4 m. It was pushed forward by the LENNY waves for several meters leaving two fresh scratch marks on the dark surface of the last interglacial reef terrace.

The exceptional impact of hurricane LENNY is shown also in the destruction of the buildings on the Slagbaai barrier in the north west, founded in 1868, which have been destroyed for the first time in history (Fig. 65). Summarized, the LENNY hurricane waves were limited in their destructive effects yet able to add several kilometers of fresh beach ridges of coarse material to the normally sheltered shorelines of Klein Bonaire and the west facing coastline of Bonaire (Fig. 70). Along the exposed eastern coast, however, which exhibits a cliff (with notch or bench) 4 to 6 m high, hurricane LENNY was not able to leave imprints except of little pieces of rock (maximum of several kg) broken here and there out of the rock pool belt.

3.4 Event of 1877

15 Hurricanes between the years 1605 and 1993 came within 75 nm from Bonaire (NHC).

Direct observations were made on the younger hurricane effects of LENNY (November, 1999) and IVAN (September, 2004). Before LENNY in 1999, tropical storm JOAN passed south of Bonaire on October, 16th, 1988, with heavy rain for days but winds of only 65 km/h. Tropical storm BRET (August 8th, 1963), was another one passing in the south, but hurricane HAZEL of October, 7th, 1954, passing about 50 nm in the north, has impacted the island most with heavy rains (125 mm within 48

Fig. 64: North of Slagbaai hurricane LENNY has formed a new rubble spit about 100 m long.

Chapter 3: Storms and Hurricane History

Pre-LENNY (1996)

Post-LENNY (2001)

Fig. 65. At Boka Slagbaai houses from 1868 have been destroyed by *LENNY* waves.

Pre-LENNY (1996)

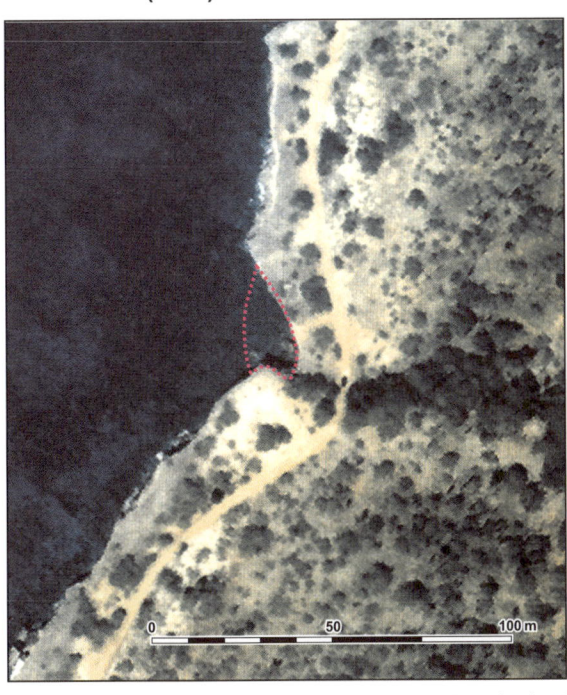

Post-LENNY (2001)

Fig. 66: *LENNY* waves have filled small incisions in the 5 m high cliff with coral rubble.

Fig. 67:
The hard contrast of very dark coral debris with cyanobacterial coating and fresh LENNY ridges can be seen at many places in NW Bonaire.

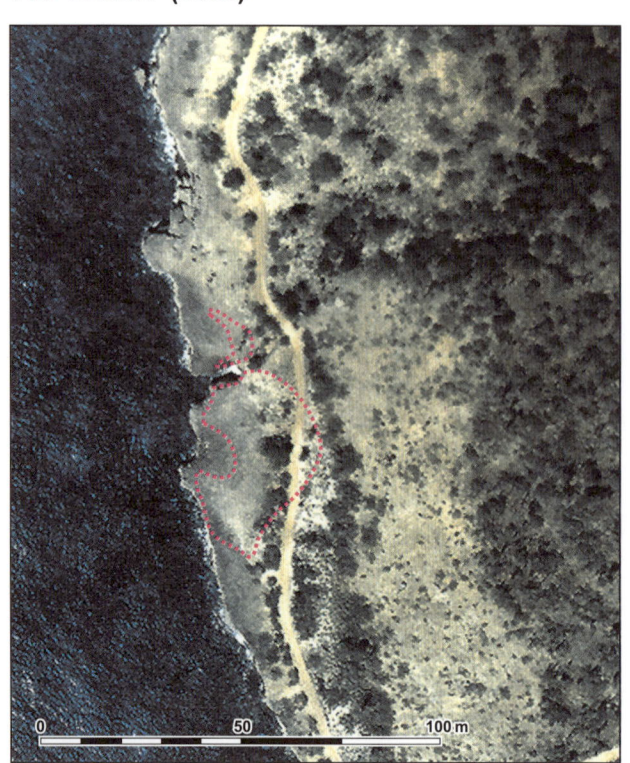

Pre-LENNY (1996)　　　　**Post-LENNY (2001)**

Fig. 68: Semicircular debris pattern from LENNY waves on top of an old coral reef terrace at about +5 m north of Slagbaai.

hours). Observations on hurricane TECLA are scarce, with no known wind speeds, but newspaper sources of September, 23th 1877, revealed 70 casualties for Bonaire. It passed the island in the south, which explains the storm formations along the mostly sheltered side of the island. Indeed a storm ridge or terrace 1-3 m wide and older than that of LENNY, but still with a limited weathering can be found up to +2 m (i.e. double as high as that from hurricane LENNY) along some parts of the leeward coast of Bonaire (Figs. 71-73). We sampled these pre-LENNY ridges at Nukowe Bay for ESR dating. This gave us a date of deposition (dated in the laboratory of U. RADTKE by KATRIN STABEN, Köln University) of 142 ± 5 years, which is in agreement with the news paper article dates.

Chapter 3: Storms and Hurricane History

Fig. 69:
Coral debris is distributed in the coastal vegetation, and some larger boulders have been broken from the cliff by *LENNY* waves in 1999.

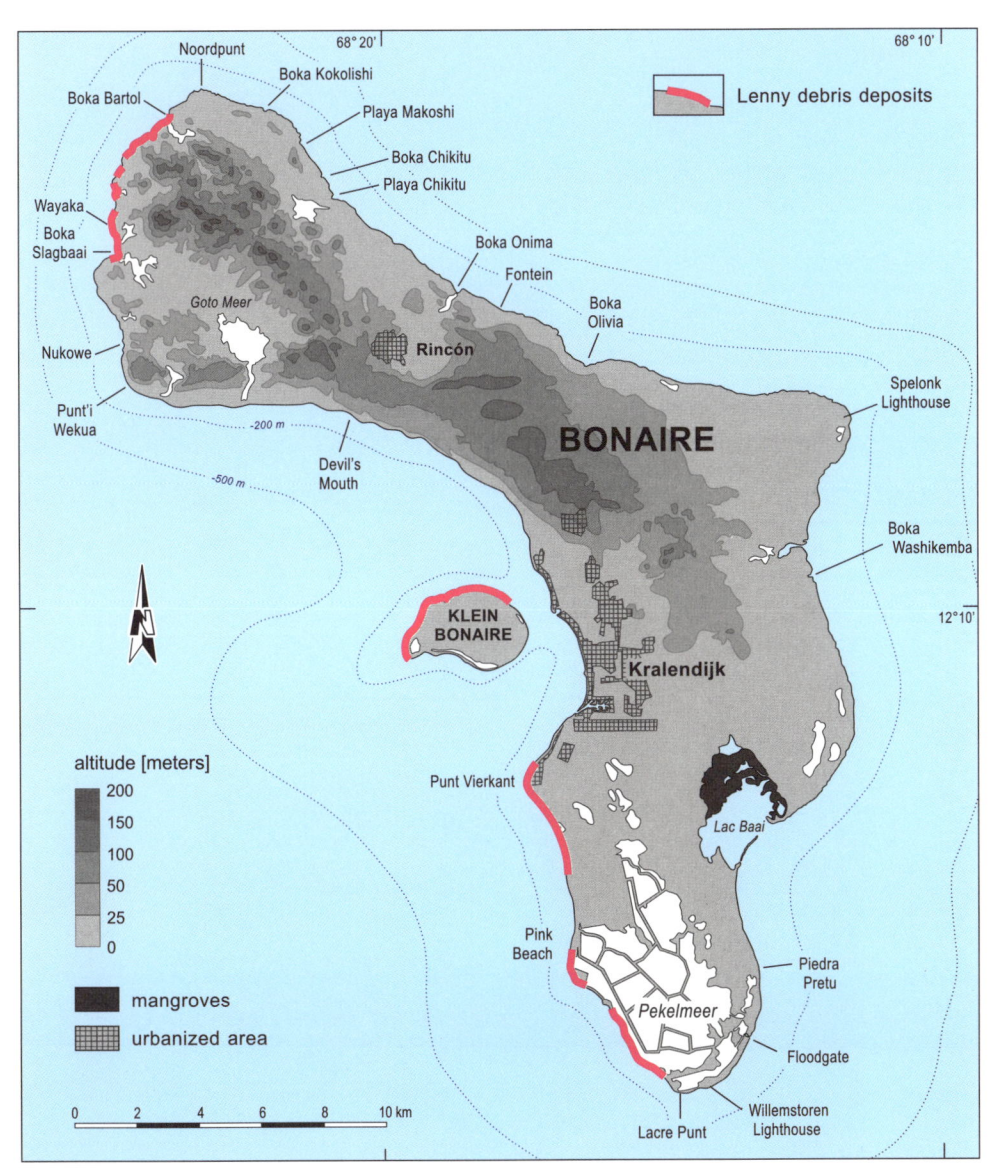

Fig. 70:
The sketch shows the areas impacted by hurricane *LENNY* on Bonaire island.

47

COASTAL RESPONSE TO EXTREME WAVE EVENTS

Fig. 71:
A small intermediate terrace of coral debris between the light *LENNY* terrace of 1999 and the top is the result of a storm from the year 1877. NW coast of Bonaire.

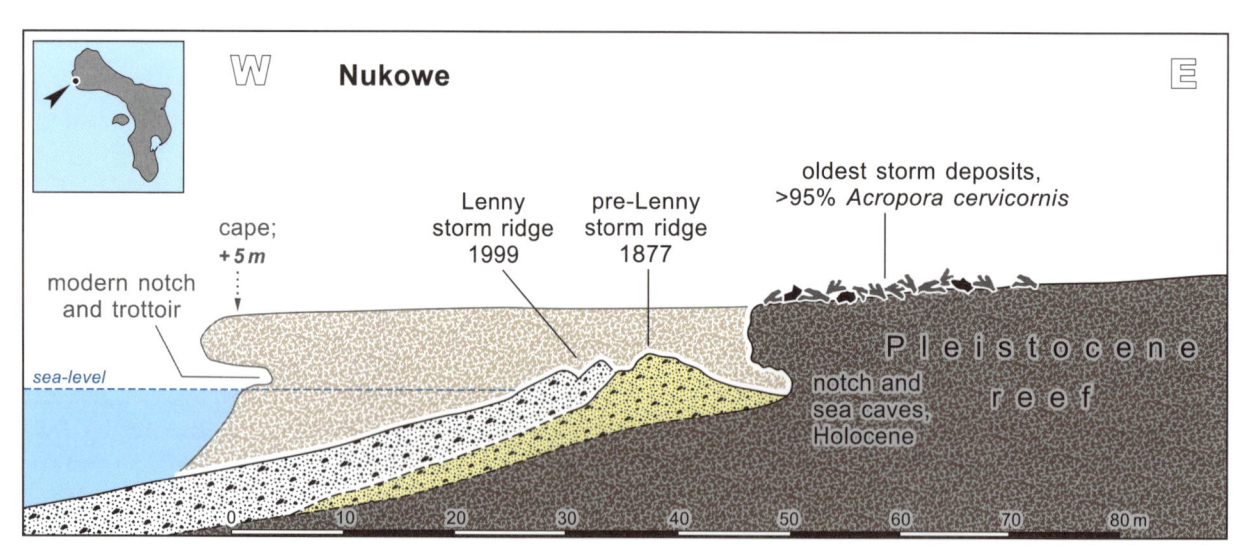

Fig. 72:
In Nukowe Bay the storm ridge of 1877 can clearly be distinguished from younger deposits.

Fig. 73:
Darker colours on these tonguelike ridge near Goto Meer show an age of many decades for the event as does the vegetation on it. The forms are the result of a storm in 1877. East of Goto Meer.

3.5 Older Storm Relics

The criteria for differentiation between storm- and tsunami deposits is under debate, yet both events have a very different geomorphological and sedimentological signature in the coastal landscape due to the difference in wave characteristics. Distinguishing between storm and tsunami contain still many pitfalls when using fine sediments – many diagnostic criteria have been forwarded in global literature to distinguish between tsunami and storm deposited fine sediments:

1. Stratigraphic characteristics of tsunami deposits are: The event unit fines upward and inland (FOSTER, 1991; DAWSON 1994). The lower contact is unconformable or erosional (DAWSON, 1988; MOORE, 1988). The deposit can contain intraclasts of reworked material (DAWSON, 1994; MOORE, 1994; GOFF, 2001; KORTEKAAS, 2002). The stratigraphic criterion reflects the high energy of such events.

2. The degree of sorting identified in the deposit. Storm deposits are generally composed of poorly to moderately sorted material with a unimodal particle size distribution (SEDGEWICK & DAVIS, 2003), whereas tsunami deposits are often reported to be composed of poorly sorted sand, clasts and debris often showing a bimodal particle size distribution (SHI et al., 1995; GOFF et al., 2004):

3. Graded beds and evidence of bidirectional flow. Many authors have noted that storm deposits contain numerous thin graded beds that are the result of individual waves during a storm washover event (LEATHERMAN & WILLIAMS, 1977; SEDGEWICK & DAVIS, 2003; NOTT, 2004; TUTTLE et al., 2004). The model of SEDGEWICK & DAVIS (1977) shows that storm beds are recorded as thin horizontal laminae in terrestrial environments and foreset beds in subaqueous systems.

4. The granulometric criteria to identify tsunami or storm deposit are: An anomalous sand unit in, for instance, a lagoon sequence (OTA, 1985; MOORE, 1988; MINOURA, 1994; DAWSON, 1994). However, the size of the particles depends on the material available on the coastline.

5. The palaeontological criteria are: An increase in abundance of marine to brackish water diatoms (DAWSON, 1988; MINOURA, 1994; HEMPHILL-HALLEY, 1996; CLAGUE, 1999). Marked changes in foraminifera and other microfossils assemblages (PATTERSON, 1996; SHENNAN, 1996; CLAGUE, 1999). Shell-richness of the event unit.

6. The geochemical criteria are: Increases in sodium, calcium, magnesium and chlorine relative to under and overlying layers (MINOURA, 1994; GOFF, 1998).

7. Dating tsunami and storm fine sediment deposits is complicated due to older debris that are incorporated in the event deposit. OSL is the method that has achieved better results. The criterion presented reflects both the high energy of the events, as well as a marine invasion. However, both tsunamis and storm surges can cause such impacts in the coastal stratigraphy. The uniqueness of a tsunami event, when compared with the more frequent storm surges, is still one of the main, if not the decisive, criteria, although not a scientifically sound one. Obviously this can only be applied in regions not subject to frequent tsunamis. It is undoubtedly important to be able to distinguish such sort of events in order to establish patterns of coastal flooding.

Therefore one aim of our field studies was to acquire adequate quantitative and qualitative data sets to distinguish between the two different phenomena.

For coarse debris deposits it is easier to distinguish between the two different events, yet it is clear that a full story on Holocene storms can only be written if all remains have been preserved at a certain location. This requires a diminishing intensity of these storms (in order that older deposits are not moved or washed away by a following storm). Furthermore, a rising sea level inhibits locating these deposits and reduces the change to establish a full storm history. In fact, the oldest storm relics we could identify on Bonaire are superimposed on huge tsunami boulders with an absolute deposition age of approximately 1300 BP (Figs. 74 and 75).

Millions of fragments of *Acropora cervicornis* can be found along the west and south west facing coasts of the northern part of Bonaire from Boca Frans to Karpata (Figs. 76-79). These fragments are sorted in size classes and the deposit lacks finer material. This is in striking contrast to the underlying bimodal tsunami deposits. These deposits also show *Acropora cervicornis* fragments but are mixed with sand and boulders up to several tons (Figs. 77 and 78). The coarse storm deposits are rather stable, whereas the tsunami deposits are characterized by an unstable framework. The height (asl) of these ridges at the leeward coasts of Bonaire reach +3.5 m near Boka Frans and Goto Meer, and at least +5 m further north (Nukowe) and south. Sometimes older giant *Acropora*

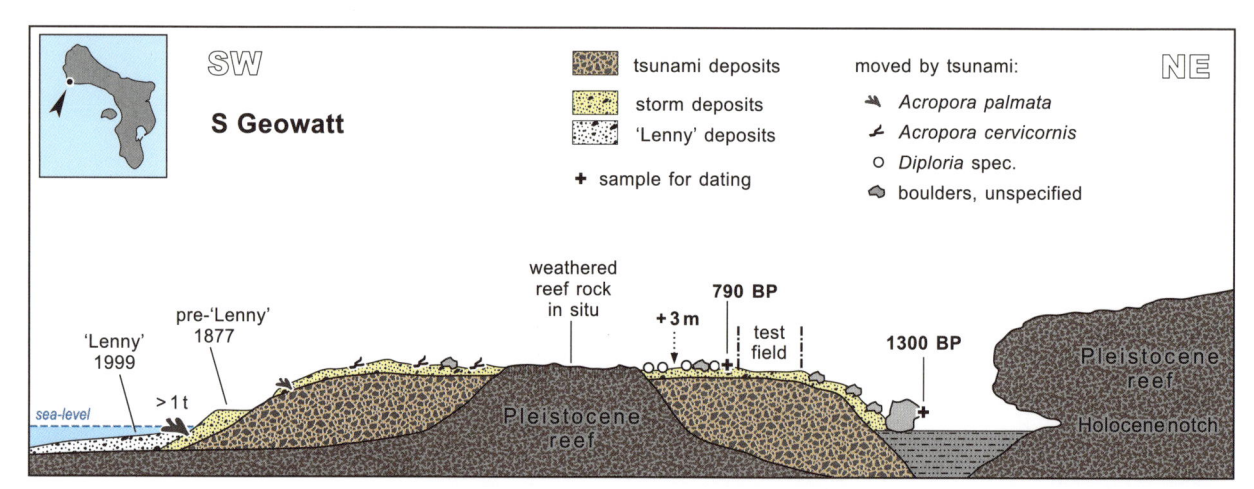

Fig. 74:
The oldest storm deposits of about 790 BP are preserved on top of an older tsunami deposit of about 1300 BP near Salina Tern.

Fig. 75:
Old large tsunami boulders (1300 BP) have been overthrown by younger storm debris (from 900 BP) near Salina Tern. See also Fig. 74.

palmata are incorporated in the older storm deposits but were not moved during younger events (Fig. 80). Along the exposed (eastern and northern) shorelines of Bonaire we were not able to detect these kind of deposits, probably because of the overwhelming existence of very coarse tsunami boulders here, but most likely because storms could not transport this debris over the cliffs of 4-6 m height along these shorelines.

Nearly all corals in this storm deposit show sharp fragmentation and are well preserved (with the species-specific sculpture intact). This may be a hint to an extreme destruction of a flourishing *Acropora cervicornis* reef which could have been typical for the leeward side of the Netherlands Antilles. As other species are nearly absent this reef could have been built by more than 95% of *Acropora cervicornis*. The huge amount of debris is an argument for a longer period of undisturbed reef growth along these coasts before the destruction. We counted 170 - 400 fragments per m², which means that this deposition layer (5-15 m land inwards) with a height of 40 cm contains approximately 10,000 to 25,000 fragments per meter (which have been broken from every running meter of the fringing reef). That equals an amount of roughly 10 tons of coral debris per running meter of coastline. The destruction of this reef therefore must have been extreme, and the assumption arises since that event a reef dominated by this species has never been developed again. But we cannot exclude, that it was reestablished between the tsunami of 1300 BP with a similar amount of *Acropora cervicornis* fragmentation, and this storm.

Absolute dating of first samples from these storm deposits gave a conventional radiocarbon age of 790 BP, which means a recovery period of 500. Comparing the depositions of this storm with the

Chapter 3: Storms and Hurricane History

Fig. 76:
Typical aspect of the surface of the oldest storm ridge at about 3-3.5 m asl along the NW coast of Bonaire with a lot of small branches from *Acropora cervicornis*.

Fig. 77:
The upper half meter of debris on top of the bimodal tsunami sediments near Karpata are from a storm about 900 years ago.

Fig. 78:
The base of this sections shows bimodal tsunami deposits (mostly *Acropora cervicornis*), whereas the upper layer is a sand free storm deposit. See also Fig. 77.

COASTAL RESPONSE TO EXTREME WAVE EVENTS

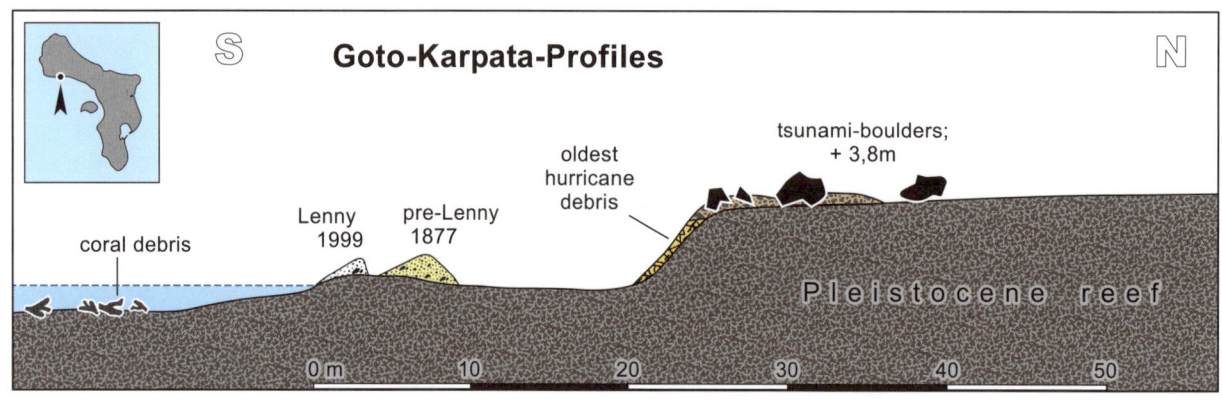

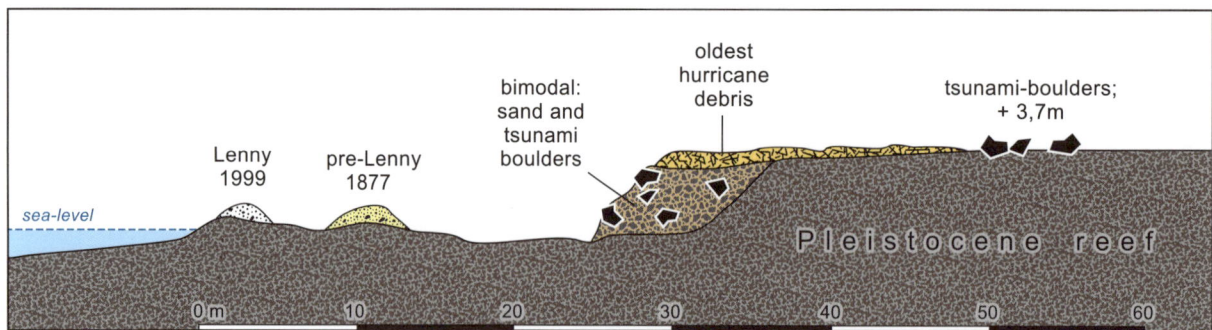

Goto-Karpata-Profiles

Fig. 79:
Between Karpata and Goto three storm deposits (*LENNY* of 1999, pre-*LENNY* of 1877, and a very old one, probably from 790 BP and up to +3.5 m high) can be seen in two different sections.

Fig. 80:
During former centuries huge *Acropora palmata* must have lived along the NW coast of Bonaire, now incorporated in older tsunami and storm sediments.

category 4 to 5 hurricanes *LENNY* and *IVAN* depositions (the first coming from the west and the latter from the east, both passing approximately 150 km north of Bonaire), it is a reasonable assumption that the storm of 790 BP with its sedimentological effects along the leeward side of the island may have been a hurricane passing the island at its southern side. This (waves pushed up due to the confinement of the narrow channel between the coasts of Bonaire and Venezuela) would explain the extreme wave power on this side, several times greater than of strong hurricanes passing the north side.

3.6. Hurricane Impacts on the Reefs of Bonaire

Studies focusing on the effects of hurricanes and tropical storms on coral reefs show that damage can be severe (STODDART, 1962) and reefs can become locally stripped from coral colonies and other organisms. Normal recovery from extreme storm damage is in the order of several decades (DOLLAR & TRIBBLE, 1993; DONE & POTTS, 1992) and depends on the severity of the damage (CONNELL, 1997). Hurricanes are also mentioned as a significant factor in speeding up reef degradation on reefs already suffering from anthropogenic induced decline (GARDNER et al., 2005). Nevertheless, strong storms not only have a negative influence on the reefs, but clearance of dominant species and loose sediments can promote biodiversity and make space for settlement of new recruits. On Bonaire the effects of hurricanes on reef ecology have not been researched, there are merely damage descriptions:

Hurricane *LENNY* has cleared to a certain amount the reef flats along the leeward side from loose sediments (building a wider beach ridge). Reef damage surveys at 33 sites conducted just months after the storm on Curacao, documented occurrences of severe toppling, fragmentation, tissue damage, bleaching, and smothering due to the storm. Several factors influenced the degree of damage experienced by the reef, including the trend of the shoreline, coral growth form, colony size, and water depth (BRIES et al., 2001). *LENNY* had the same effect on Bonaire (pers comm. SR Scheffers).

Hurricane *IVAN*, coming from the north has not much affected the reefs in the south except for clearance of loose sediments as seen with hurricane Lenny. Some sites experienced damage to the fragile *Acropora cervicornis* and *Acropora palmata* stands (Fig. 81). Most corals in the south, however, were covered with very fine sand, but this was removed after days, either by wave-action or mucus-shedding of the corals (Fig. 82). One site, Baby Beach, just south of Lac Baai was totally destroyed by waves of this severe hurricane (SCHEFFERS et al., 2005). Here, all smaller and larger corals were toppled, broken, and transported down slope. The reason for this hugely localized destruction is unknown.

Fig. 81:
Disturbed and living *Acropora cervicornis* within a sandy environment on a shallow reef flat at the west coast of Bonaire. Water depth approximately 5 m.

Fig. 82:
Head corals (*Diploria* sp.) can better survive storm impacts on the shallow fringing reefs along the leeward side of Bonaire. Water depth approximately 5 m.

4. Tsunami Imprints on Bonaire

Although more than 2,000 publications exist on the tsunami phenomenon worldwide (compare e.g. BRYANT, 2001; DAWSON, 1994, 1996; DAWSON et al., 1991; GOFF et al., 2001; HEARTY, 1997; KELLETAT 2003; KELLETAT & SCHELLMANN, 2001, 2002; KELLETAT & SCHEFFERS, 2004a,b,c; KELLETAT et al., 2005; MASTRONUZZI & SANSO, 2000; MINOURA & NAKATA, 1994; NANAYAMA et al., 2000; NISHIMURA & MIYAJI, 1995; NOTT, 1997, 2000, 2003a,b, 2004; NOTT & BRYANT, 2003; SATO et al., 1995; SCHEFFERS, 2002a,b, 2004, 2005a; SCHEFFERS & KELLETAT, 2003; SCHEFFERS et al., 2005; SHI et al., 1995; WHELAN & KELLETAT, 2002, 2003a,b; ZAHIBO & PELINOVSKY, 2001), only 4% of all tsunami related papers deal with geomorphologic or sedimentologic tsunami imprints. This is the reason that, in this paper, we focus exclusively on these imprints. We emphasize on Bonaire, to demonstrate the variety of tsunami proofs on this single spot in the Caribbean, although other regions in the world definitely show paleo-tsunami relics as well.

4.1 Historical Sources

LANDER & WHITESIDE (1997) as well as LANDER, WHITESIDE & LOCKRIDGE (2002) have compiled all historical sources on possible tsunami events during the last 500 years in the Caribbean. Out of 88 tsunami descriptions, they mention 27 reliable dates for significant tsunami, which caused destruction or other remarkable imprints including fatalities. Table 3 summarizes the events with the highest run up values. The oldest event is of the year 1530 from the Cariaco Bay on the north coast of Venezuela. Even the 1755 Lisbon tsunami reached the Caribbean with run up values of several meters high. Younger events are the strong tsunami of 1867 from the Virgin Islands, and the 1918 event on the northwest coast of Puerto Rico (see also Fig. 3). No tsunami, however, is mentioned for Bonaire or the neighboring islands, and no field evidence has ever been published for tsunami relics before the year 2002 (SCHEFFERS, 2002a). From other parts of the Caribbean there is information on tsunami deposits (compare MOYA (1999) for the Puerto Rico event of 1918, from JONES & HUNTER (1992) for Grand Cayman, from SCHUBERT (1994) for Venezuela, WEISS (1979) for an island north of Venezuela, or from TAGGART et al. (1993) for Isla de Mona west of Puerto Rico). These events have been dated to historical times or slightly older up to 1300 BP. All sources, such as historical reports,

Table 3: Historical run up heights in the Caribbean (extracted from LANDER & WHITESIDE, 1997).

Year	Region affected	Maximum run up [m]
1530	N Venezuela	7.3
1755	Saba	7.0
1780	St. Martin	4.5
1842	Jamaica	3.0
1856	Guadeloupe	8.3
1867	Honduras	5.0
	Virgin Islands	7.6
1918	Guadeloupe	10.0
1946	Puerto Rico	6.0
	Dominican Republic	2.5

field evidence, registrations of tide gauges and the geodynamic situation of plate boundaries with active volcanoes, show that tsunami of significant power are possible and have happened in the Caribbean area, and certainly will happen again.

4.2 Tsunami Sediments on Bonaire

Fine tsunami sediments maybe interpreted as storm deposits, and coarse debris including large boulders have been mostly identified as being the result of extreme hurricane waves. So far, no proved method exists to differentiate storm debris from tsunami boulders. In chapter 3 we have presented our observations on the depositional and geomorphological power of the two extreme hurricanes *LENNY* (1999) and *IVAN* (2004), and in particular their limited forces to transport large boulders. The physics of boulder movement is elaborated by NOTT (1997, 2003a,b). The authors' discussion of the transport power of probably the most powerful storms on earth (NOTT, 2004) also shows that boulder transport by storm waves along shoreline reaches a limit close to 10 or 20 tons above which no boulders can be moved (mostly far below these values). Our calculations of a transport figure (weight of a boulder in metric tons, multiplied with the vertical transport (against gravity) and distance to the shoreline in meters) resulted in figures higher than 2,000 definitely belong to tsunami waves. Storm waves also leave a different pattern of boulder deposition than tsunami. The latter produce good imbrication of huge clasts, vertical position of fragments, unstable setting, and in particular mixing with finer particles like sand to show a bimodal deposit with a chaotic structure and boulders or clasts floating in sand. Storm waves do not show these patterns. Here we would shortly like to demonstrate the different types of coarse tsunami deposits which can be found on Bonaire.

A first impression is that tsunami deposits can be found far out of the reach of strong storm waves, with respect to altitude above sea level and distance from the surf belt. Another impression is the huge amount of material in a single depositional unit (Figs. 83-90). The most widespread tsunami phenomena are broad ridges of boulders from 5 to more than 100 kg along the south coast east of Willemstoren lighthouse, in the northwest of Bonaire, and around Klein Bonaire. Partly large boulders of several tons may be incorporated (Figs 91 and 92).

These ridges can be up to 3 m thick, far more than 50 m wide and several kilometers long (Figs. 84 and 86). The original bimodal structure (boulders

Fig. 83:
Coral debris deposit near Salina Tern: the upper section from a storm about 900 years ago, the lower one, separated by a thin soil layer, shows sand, rubble and boulders from a tsunami event about 1300 years ago.

Fig. 84:
Remnant of a tsunami ridge up to 3 m high, some kilometers long and originally 50-80 m wide along the south coast of Bonaire, destroyed by mining along the landward side (left in picture). In the front part fresh erosion by *IVAN* hurricane waves.

COASTAL RESPONSE TO EXTREME WAVE EVENTS

Fig. 85:
Coral debris in a tsunami ridge about 500 years old from the south coast of Bonaire: fresh broken coral mixed with rounded rubble.

Fig. 86:
A wide tsunami rampart, typical for the east coast of Bonaire, at around +5-6 m and up to 400 m inland. Fragments mostly broken off from a cliff and up to more than 200 kg. Age is from about 500 years up to more than 4000 BP.

Fig. 87:
Only the base of tsunami ridges and ramparts show the sandy matrix, whereas the upper parts have been washed out by heavy rain and storm splash.

56

Chapter 4: Tsunami Imprints

Fig. 88:
The tsunami rampart with dark colours on the debris by cyanobacteria etc. as a ridge up to 3.3 m high and 90 m wide in front of Boka Bartol. Young hurricane waves have only washed on a limited amount of smaller debris.

Fig. 89:
Aerial photograph (1996) of the Boka Bartol tsunami barrier.

Fig. 90: Aerial photograph of the Salina Tern barrier (1996).

and sand) has been changed to a sand free upper layer of 0.5-1 m and fresh looking sand, shell and boulders in the basal parts (Fig. 87). Although abraded for more than 1-2 m by hurricane surf they represent an amount of sediment which is at least 100 times greater than that of one single storm.

Along the east coast of Bonaire, the tsunami debris forms a rampart like feature: 0.5 to 5 m thick, 30 to 100 m wide and many kilometers long. These ramparts cover the slightly uplifted coral reef terrace of the last interglacial with a sediment free belt of rock between these deposits and the cliff. These deposits are called ramparts by FOCKE (1978a,b,c) and others. The particles are mostly larger than those of the ridges in the south, weighing on average 50-200 kg (Fig. 86) and many fragments weigh 20 or more tons. These ramparts contain millions of tons of limestone debris from interglacial reefs. Much of these deposits, however, have been removed and mined for building purposes, landfill for Bonaire's airport, and other reasons, and only the biggest boulders have been left *in situ*. This mining exposed the inner character of the ramparts, which is similar to that of the tsunami ridges: it is of a bimodal nature with sand and boulders mixed chaotically and with the upper layer sand washed out. The debris of the boulder ridges and ramparts all show a darkgrey coating due to the presence of endolithic organisms such as cyanobacteria and chlorophyceae. Sediments of known age along Bonaires coasts such as storm ridges (back to at least 1877 AD) exhibit the same feature. Thus, the color represents an age of at least 100 years. If this dark color is found many decimeters deep within the deposit, the ages should be more than 100 years old and during this time no disturbance took place, which argues for extremely rare and extremely energetic events. These might be exceptional 500 or 1000-year storms or tsunami. An important argument for a tsunamigenic origin is the incorporation of giant boulders into these ridges and ramparts (>50 tons) away from the surf zone, which have the same age as the ridges or ramparts. Absolute dating shows ages from approximately 500 BP to more than 4000 BP (see chapter 4.4).

Typical is the asymmetry of tsunami ridges and ramparts, in particular along the east coast of Bonaire (Fig. 93): The seaward part has been abraded and truncated by the strongest hurricane waves, as could be observed by hurricane *IVAN* in

Chapter 4: Tsunami Imprints

Fig. 91:
Fragment of weathered *Acropora palmata* with a weight of about three tons as a remnant of a tsunami deposit, excavated by hurricane waves at the northwest coast of Bonaire.

Fig. 92:
Some ramparts show imbricated boulders with several tons of weight, here partly excavated from finer debris by hurricane *IVAN* waves. Coast NE of Lac Baai.

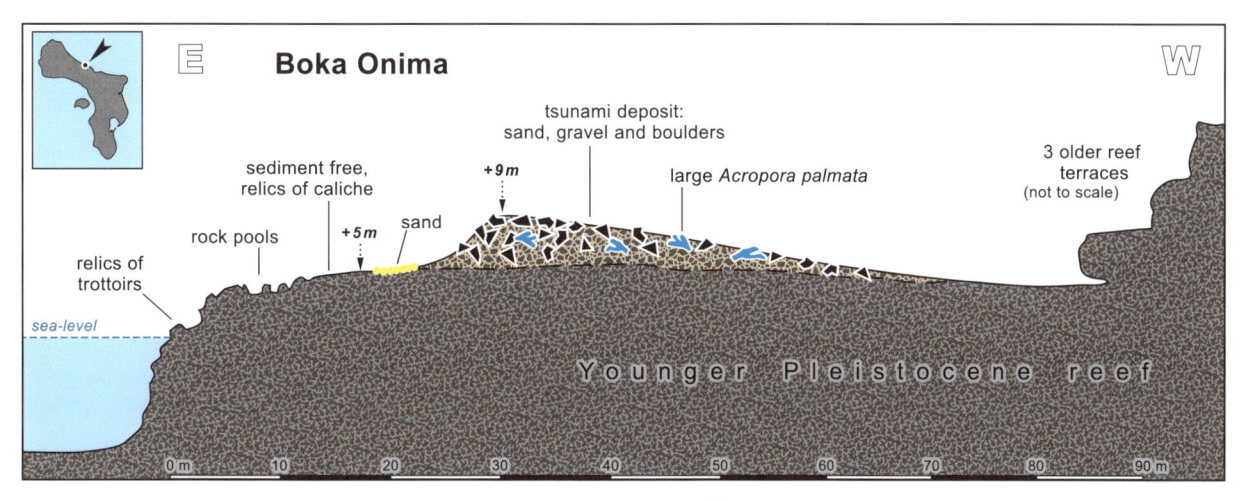

Fig. 93: Remnant of a tsunami rampart at Boka Onima, east coast of Bonaire.

COASTAL RESPONSE TO EXTREME WAVE EVENTS

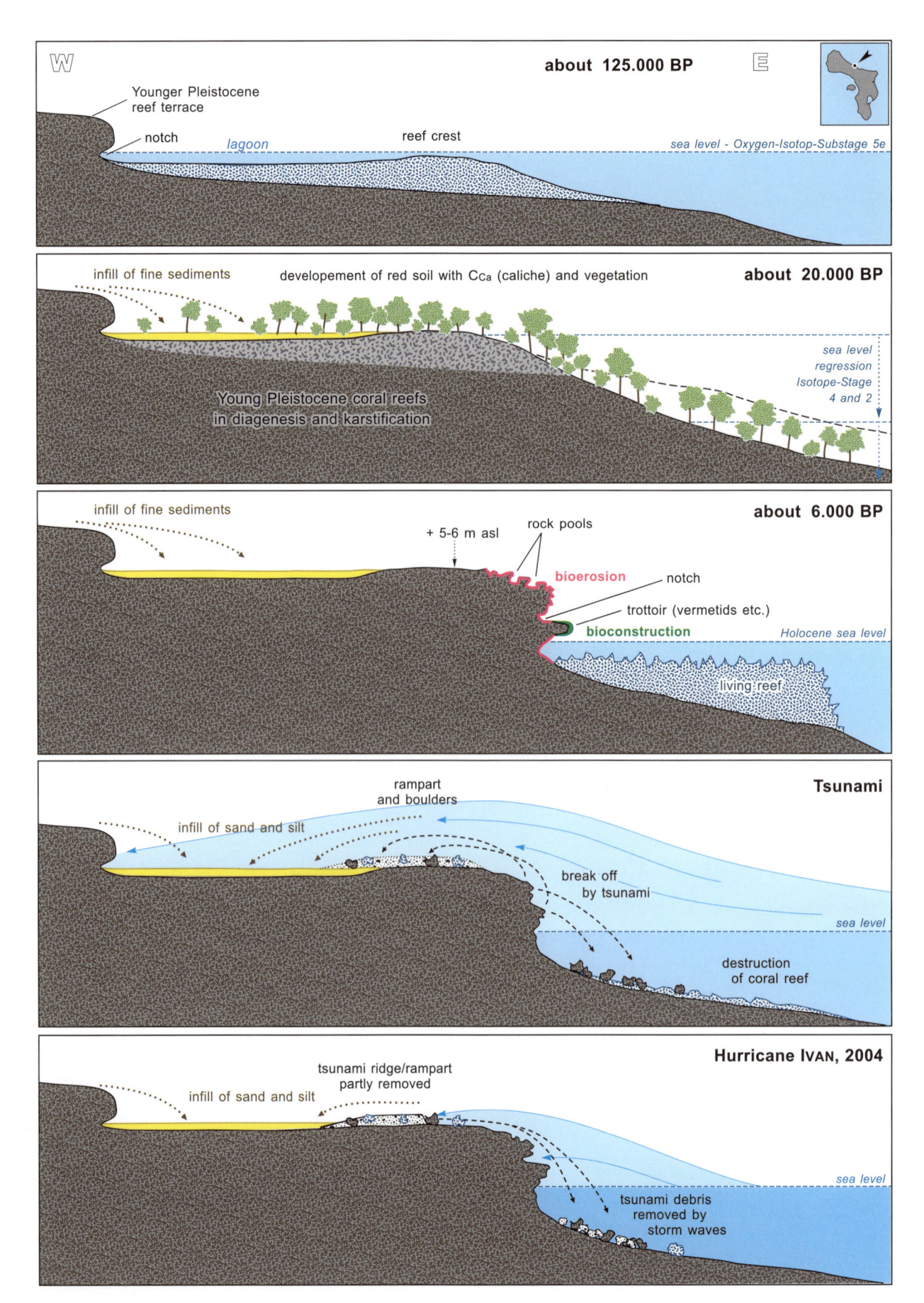

Fig. 94: The development of a Holocene tsunami rampart on the Youngest Pleistocene coral reef terrace.

Chapter 4: Tsunami Imprints

Fig. 95: Mining of tsunami debris and destruction of a rampart along the east coast of Bonaire in the Washikemba area. Oblique aerial photograph of 2001.

Fig. 96:
Tsunami boulder at Washikemba, 180 m from the cliff, with very well preserved bioerosive rock pools from the supratidal belt. The boulder has a weight of nearly 5 tons.

September, 2004. This means, that the amount of debris accumulated onshore by tsunami was much greater than can be detected now. Fig. 94 shows the development of a Holocene tsunami rampart on a Younger Pleistocene coral reef terrace. However, extensive mining of millions of tons of carbonate fragments during the last 30 years severely depleted the original tsunami debris on the reef terrace (Fig. 95).

The third kind of tsunami deposits of Bonaire are large single boulders or clusters of boulders, weighing 20 tons, 50 tons or even around 200 tons each (Figs. 96-110). They can be found – like the ramparts – at altitudes of +5 m and higher and up to 400 m away from the cliff.

These boulders represent a transport energy which is much higher than can be produced by the strongest of storm waves, reaching transport figures (weight in tons × distance transported × altitude asl) of several 100,000, i.e. more than a thousand times greater than for the largest hurricane boulders. Some

61

COASTAL RESPONSE TO EXTREME WAVE EVENTS

Fig. 97:
This boulder of about 120 tons and a length of 9.2 m can be found near Playa Chikitu. The extended sandy flats derive from that beach by trade winds and show some small dune ripplets. The seaward limit of the sand has been washed out by strong storm waves.

Fig. 98:
A cluster of smaller (1-5 tons) and larger (up to 130 tons) boulders at Seru Grandi, 6 to 8 m asl.

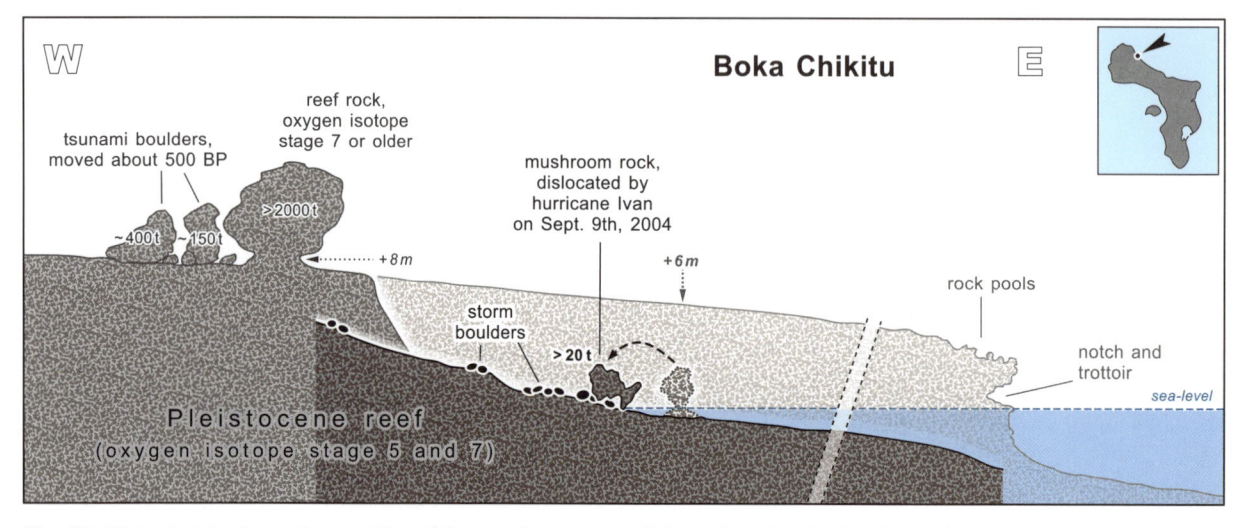

Fig. 99: This sketch shows the position of the mushroom rock (22 tons) broken by hurricane *IVAN* waves on the bottom of Boka Chikitu, and tsunami boulders of 150 and 400 tons at +8 m farther inland.

Chapter 4: Tsunami Imprints

Fig. 100:
The 400 tons tsunami boulder near Boka Chikitu.

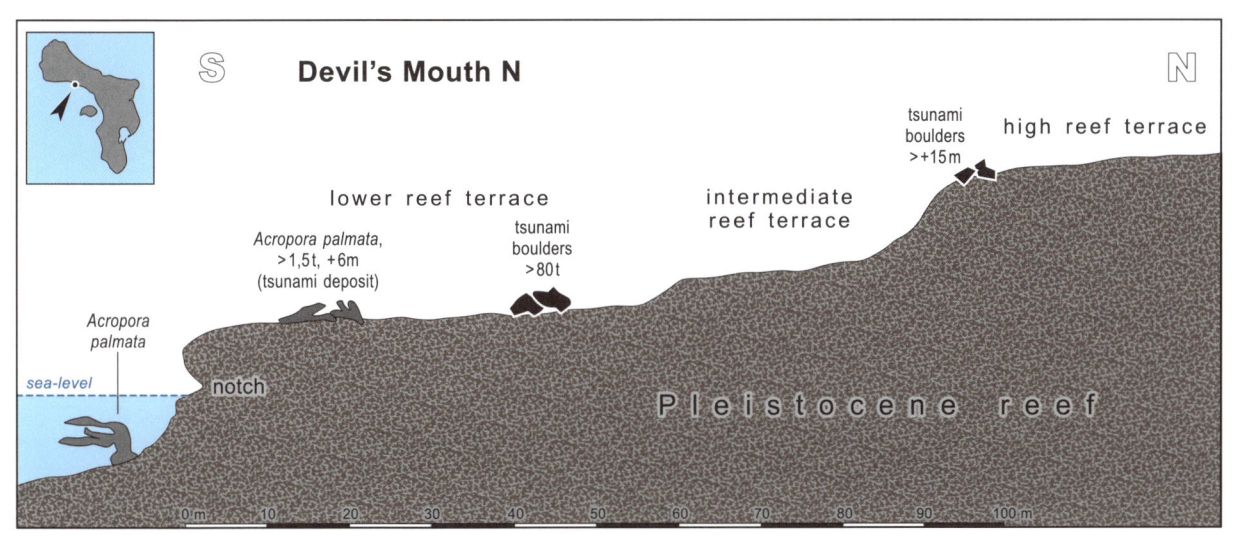

Fig. 101: Even along the leeward side of Bonaire large boulders have been dislocated by tsunami: *Acropora palmata* branches of more than 1.5 tons, or reef rock boulders up to 80 tons and smaller ones at +15 m like at Devil´s Mouth.

Fig. 102:
The largest boulder of at least 80 tons at +7 m asl in the Devil´s Mouth area of Bonaire.

COASTAL RESPONSE TO EXTREME WAVE EVENTS

Fig. 103:
At Spelonk these two pieces of rock, each with a weight of more than 120 tons, have been transported as a single boulder and broken during smash down from a tsunami wave. The site is at +5.5 m asl and 160 m from the cliff.

Fig. 104:
Mapping of large boulders from the Spelonk area. At least 89 boulders with a weight of 10 to 50 tons and 32 with more than 50 tons can be found here in a belt 100-150 m from the sea at about +5 m asl.

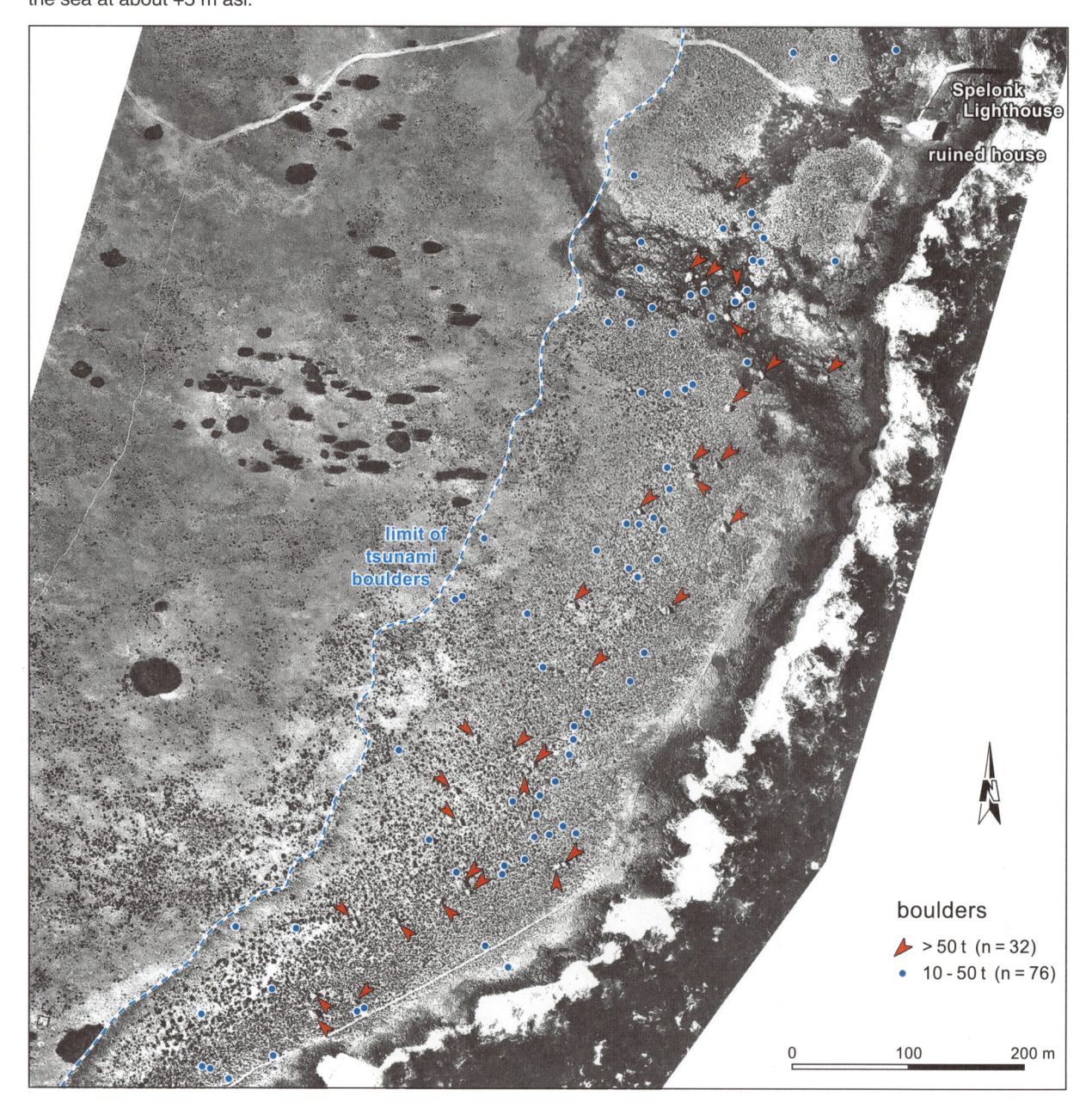

boulders can be traced to their place of origin (from the supratidal rock pool belt or from the eulittoral bench or notches, Fig. 96), and betray the kind of movement to which it was exposed (it was not a rolling along the longest axes, but a transport in a swash of water without touching the ground, smashing down with breaking of the boulders in several large pieces, Fig. 103).

The most impressive site of huge boulders clusters, many of them with weights of 50 to 150 tons, can be found near the Spelonk lighthouse on the central east coast of Bonaire island. The sequence of pictures (Figs. 104-110) shows ensembles and single boulders from this site and from the Seru Grandi area in the north. The large tsunami boulders shown here are deposited on a nearly flat surface far away from any slope or ledge. Their origin is the modern cliff side, the sublittoral or foreshore. The boulders are mixed with smaller debris – though much of this debris weighs 0.5-5 tons – and are incorporated in a rampart like feature or loose bed of smaller coral debris. This debris is well preserved and can be used for species determination.

Therefore there is no doubt that huge waves have transported this material. Difficulty to identify in the field, but identifiable from aerial photographs, are other structures backing up the argument that this material has a water born origin. These structures consist of ripple marks in finer debris or coarse boulder fields (Figs. 111-113). They all show the flow direction of the tsunami waves: from the sea perpendicular inland or with a small angle to the cliff.

Fig. 105:
Boulder field at Spelonk in dense *Conocarpus* vegetation.

Fig. 106:
Single boulder with 7 m axis, partly covered by vegetation, at Spelonk.

COASTAL RESPONSE TO EXTREME WAVE EVENTS

Fig. 107:
This large boulder shows its origin from a Pleistocene reef with dominant *Acropora cervicornis*.

Fig. 108:
Boulders of more than 100 tons at Seru Grandi. The rock platform at 7-8 m asl is cleared by hurricane waves from smaller debris.

Fig. 109:
Another aspect of the Seru Grandi tsunami boulders.

Chapter 4: Tsunami Imprints

Fig. 110:
The largest of the Seru Grandi boulders shows some vegetation on its top, while its environment has been washed out by extreme storm waves.

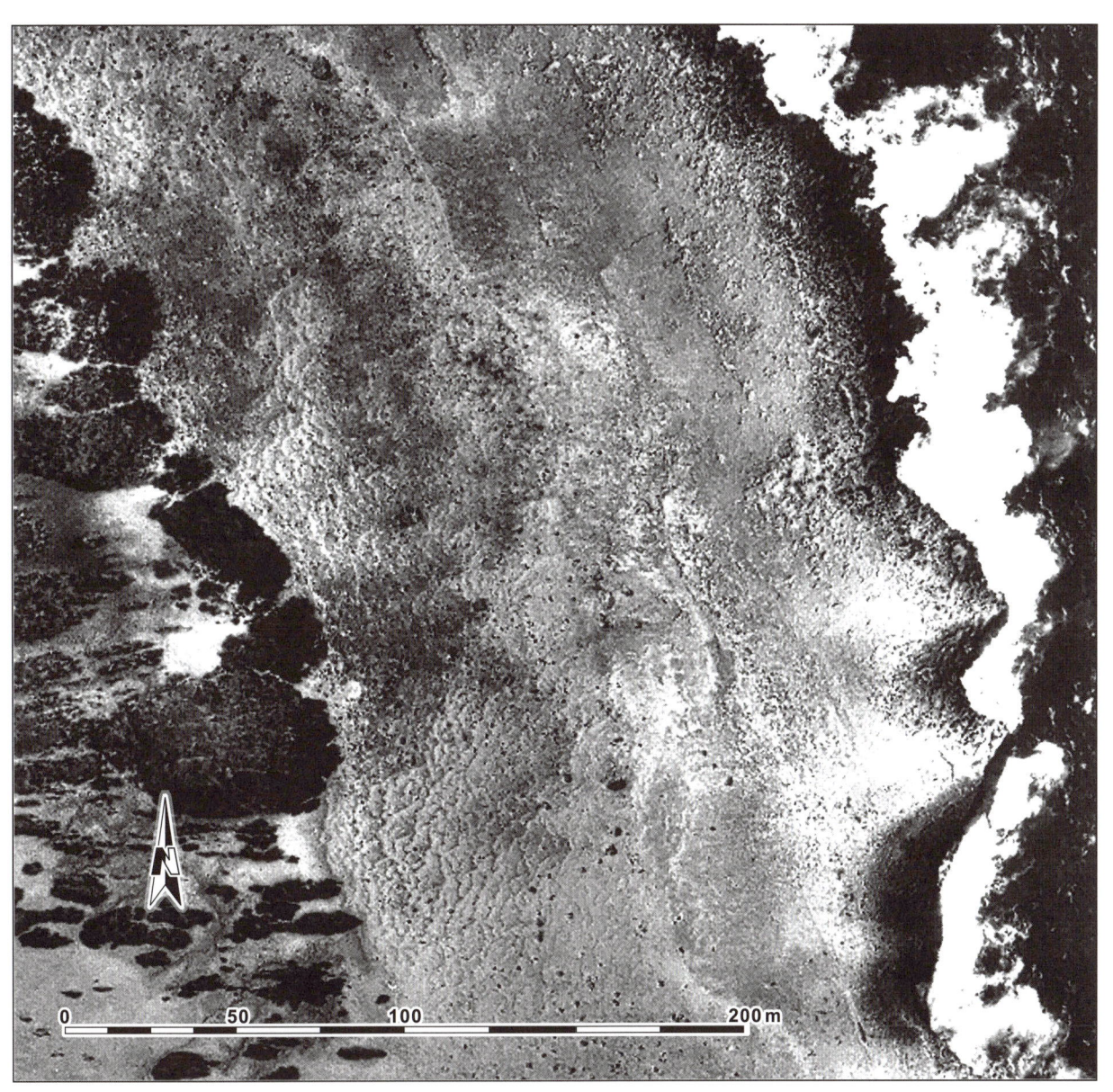

Fig. 111: Ripple marks in smaller boulder ramparts north of the Seru Grandi rock at +9 m and 100-150 m from the sea (Aerial photograph of 1961).

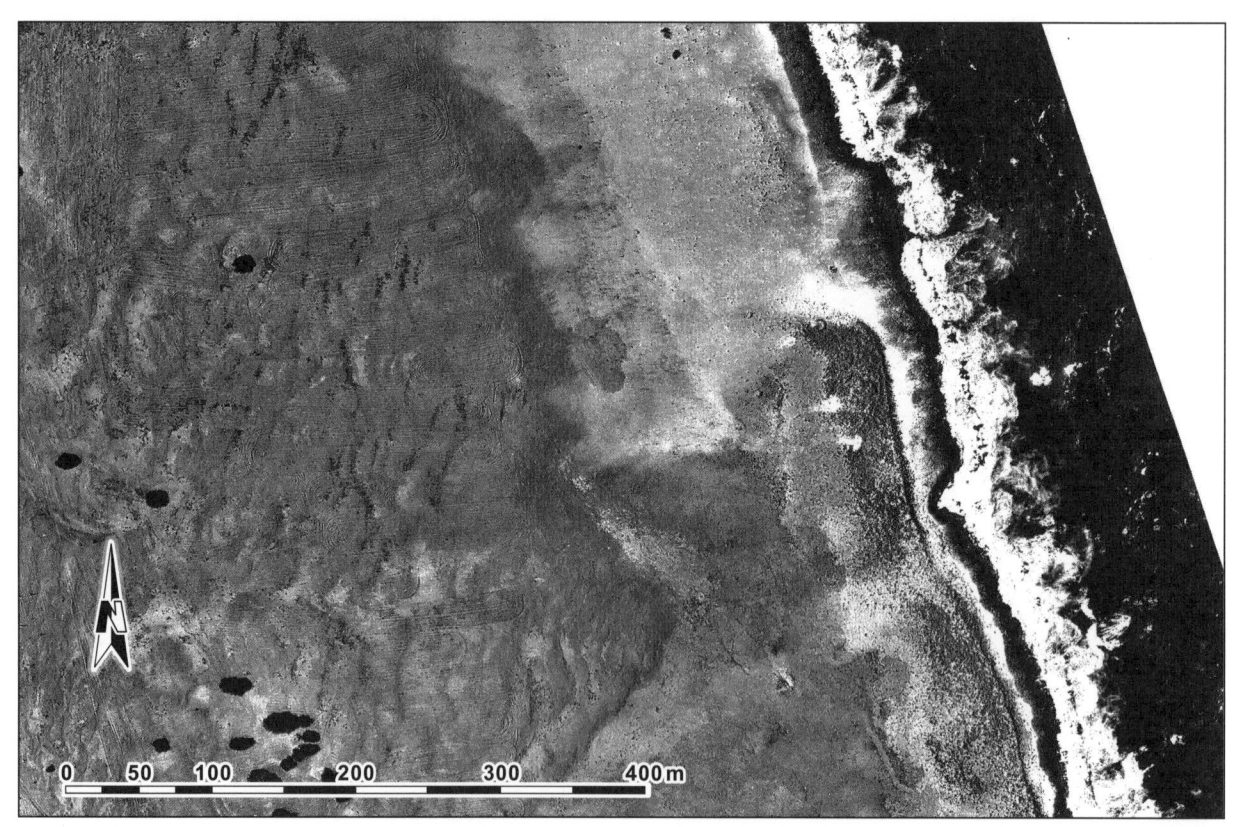

Fig. 112: This field of ripple marks is more than 200 m wide in the southern Washikemba area and up to 400 m inland at +6-7 m asl. The coarse rampart (boulders 0.5-1 ton) near the cliff shows smaller ripples, as well. Its seaward marked cliff is the result of extreme storm waves, eroding a small part of the original tsunami deposits (1961).

There are certainly much more tsunami deposits on Bonaire than described here, in particular the finer sediment structure and distribution in the wider embayments. Due to the high costs of drilling in water, this field of research remains open.

4.3 Relative Dating of Tsunami Events

Since historical sources do not mention strong tsunami for Bonaire and the neighboring islands, but tsunami relics can be found nearly everywhere, the question arises, from which times these boulder dislocations might be. An interesting hypothesis is that they are dislocated before the settlement of Europeans, i.e. before 1634 AD (Dutch), and even before the 16th century (Spaniards), because otherwise these people would have mentioned extreme events, would have suffered much from their effects, and consequently would have documented this.

These historical arguments are a first approximation for the age of the tsunami boulder dislocation. But there are more, which can be elaborated in the field: A time limit for all the tsunami boulder deposits is the age of the underlying rock, which represents the last interglacial coral reef of Oxygen Isotope Stage 5 (mostly substage 5e, see SCHELLMANN et al., 2004). Between this reef and the tsunami deposits, however, there is a long time-span of 100,000 years or more, because all the deposits described here are of Younger Holocene age, belonging to a sea level similar to today. Relative age indicators (SCHEFFERS, 2002a,b, 2003a, 2004, 2005a) comprise – amongst others –

1. Relation to historical structures of known age: the Piscadera area on Curaçao. Here, boulders from 50 to 194 tons have been deposited about 500 years ago (conventional radiocarbon data). They existed in 1634 AD, when the Dutch had built their gunpowder storage between the large boulders to protect their fortification in case of an explosion.

2. Soil and vegetation development under, beside (more precisely between the boulders and the surf area) or on the boulders, including lichens.

3. Kind and intensity of weathering like honeycombs and tafoni on carbonate sandstones (e.g. eolianites) and in particular karst solution microforms, or the dislodgement of single corals from rubble by weathering of less cemented parts in the rock. On some boulders karren features or solution pits can be observed.

Chapter 4: Tsunami Imprints

Fig. 113: In undisturbed tsunami ramparts the landward limit of tsunami can be identified by a sharp line, at which the ripple field ends. This is about 200-250 m from the sea (south part of Washikemba) (1961).

69

4. Transformation of minor forms on boulders such as parts of notches from the eulittoral or of rock pools from the supratidal. The kind of preservation may give a good idea on the relative age of transportation (compare also Fig. 96) as well as on the kind of transportation (for example in Fig. 96: a rolling-over on rough ground for more than 100 m would have destroyed the fine edges of the rock pools).

5. Transformation of cliff parts off which boulders have been broken. This will destroy all finer sculptures such as rock pools or notches and benches, and a new formation needs time. For the bioerosional process in the case of rock pools and notches the velocity of forming is in the order of 2 mm/year, i.e. rock pools of several decimeters of depth and widths of more than one meter need centuries to form again.

6. Preservation of fine outer structure of corals or shells, dislocated together with the boulders. Among all debris in ridges and ramparts most of the fragments show well rounding, boring of organisms, or covering by calcareous algae. Only a small number is very well preserved, so that the fine sculpture which determines taxon can be identified. If this is preserved in a position open to weathering this may be a sign for a young age of deposition, since the fine structure consisting of carbonates can be easily dissolved by rain.

7. Degree of washing out of finer particles from boulder clusters, ramparts or ridges, in particular the depth up to which sand has been removed from bimodal deposits.

8. Kind of affection by later storms like transforming of the seaward limit into a steeper slope, or abrading of ramparts into bowlike segments.

9. Growth of new coral on a reef affected by a tsunami and the development of a reef community after the event.

These are just the most important arguments.

Relative dating is a very important instrument, because it excludes risks incorporated in sampling and absolute dating, and it is the only way to tell the complete tsunami history. Some examples are given with the figures 114 to 120. They show differences in color, the size of material or other aspects pointing to more than one tsunami event.

The extend of tsunami deposits, i.e. of ridges, ramparts and in particular large boulders on Bonaire island is shown by the compiling map in Fig. 121.

4.4 Absolute Dating of Tsunami Events and the Tsunami Risk of Bonaire

On Bonaire and the neighboring islands as well as many other parts of the Caribbean coastal debris originates mostly from carbonates of organisms such as corals, mollusks, or calcareous algae. Therefore material can be found which is suitable for radiocarbon dating and absolute dating by ESR (Electron-Spin-Resonance), if the substance is not calcified (aragonite changes into calcite in the course of time). The aim of absolute tsunami dating, however, is not to fix the age of a dislocated rock (which in many cases is of Young Pleistocene age derived from the last one or two interglacial coral reef terraces), but to date the last tsunamigenic movement of the debris. If boulders have been dislocated from the Pleistocene cliff rock, i.e. from the supratidal environment, there is no chance for dating their movement. In most cases, however, one can find coral fragments from the living reef or mollusks out of the foreshore mixed with these boulders. The sublittoral environment not only exhibits living coral, but sediments of different types conserved from earlier storms (and tsunami) are present. To avoid mistakes with sampling, only fragments without bioerosive borings or coating of calcareous algae should be taken for dating. The catch is that sometimes very fresh looking coral can be found which gives a date much older than others at the same site, and this has to be interpreted in the right way. We encountered these anomalies, and our hypothesis is that these fragments have been buried in the sand for longer time and preserved from boring and coating.

Thorough observation may yield organisms which are attached to large boulders like coral, barnacles, oysters, vermetids, calcareous algae, and boring bivalves inside the rock. These organisms have a marine life-history and had to be killed by a sudden onshore dislocation from their living environment, and therefore are suitable to date the tsunami event.

Radiocarbon dating from littoral or marine environments raises the problem of the so-called "reservoir effect", which makes data significantly older than they really are. A time span of circa 400 years is taken as the reservoir error. Our samples showed a large number of absolute conventional radiocarbon ages of approximately 460 to 600 BP, which in fact should be at least 350-400 years old by historical facts. Examples are tsunami deposits from an extreme event not mentioned in historical sources, but containing ballast stones from European ships and dated to 500 BP (on Long Island,

Chapter 4: Tsunami Imprints

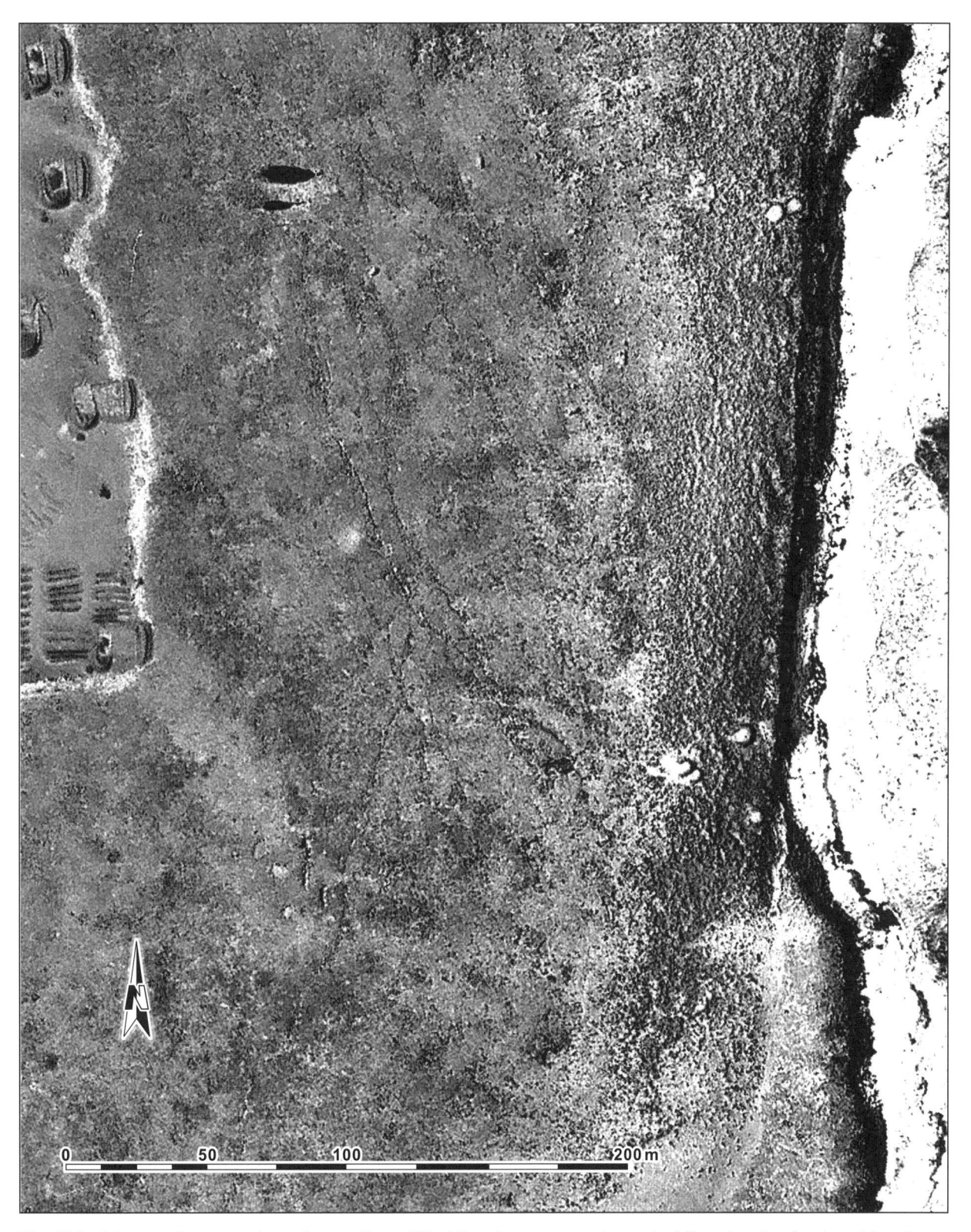

Fig. 114: A tsunami rampart from the southern Washikemba area, east coast of Bonaire: the front part has been transformed into a debris cliff by later storms (1961). The coarsest belt of fragments shows ripple marks (in debris of 0.5 to more than 1 m of length and weights of up to 1-1.5 tons!), the landward belt has a lighter color, and landward bowlike sharp ridges of coarse fragments document the reach of largest tsunami waves. All in all a document for several tsunami or several large waves of a single tsunami.

Fig. 115: This aerial photograph (from 1961) of the Washikemba area shows a pattern of different boulder sizes and colors from the cliff to inland, giving relative indices for age differences of the deposits.

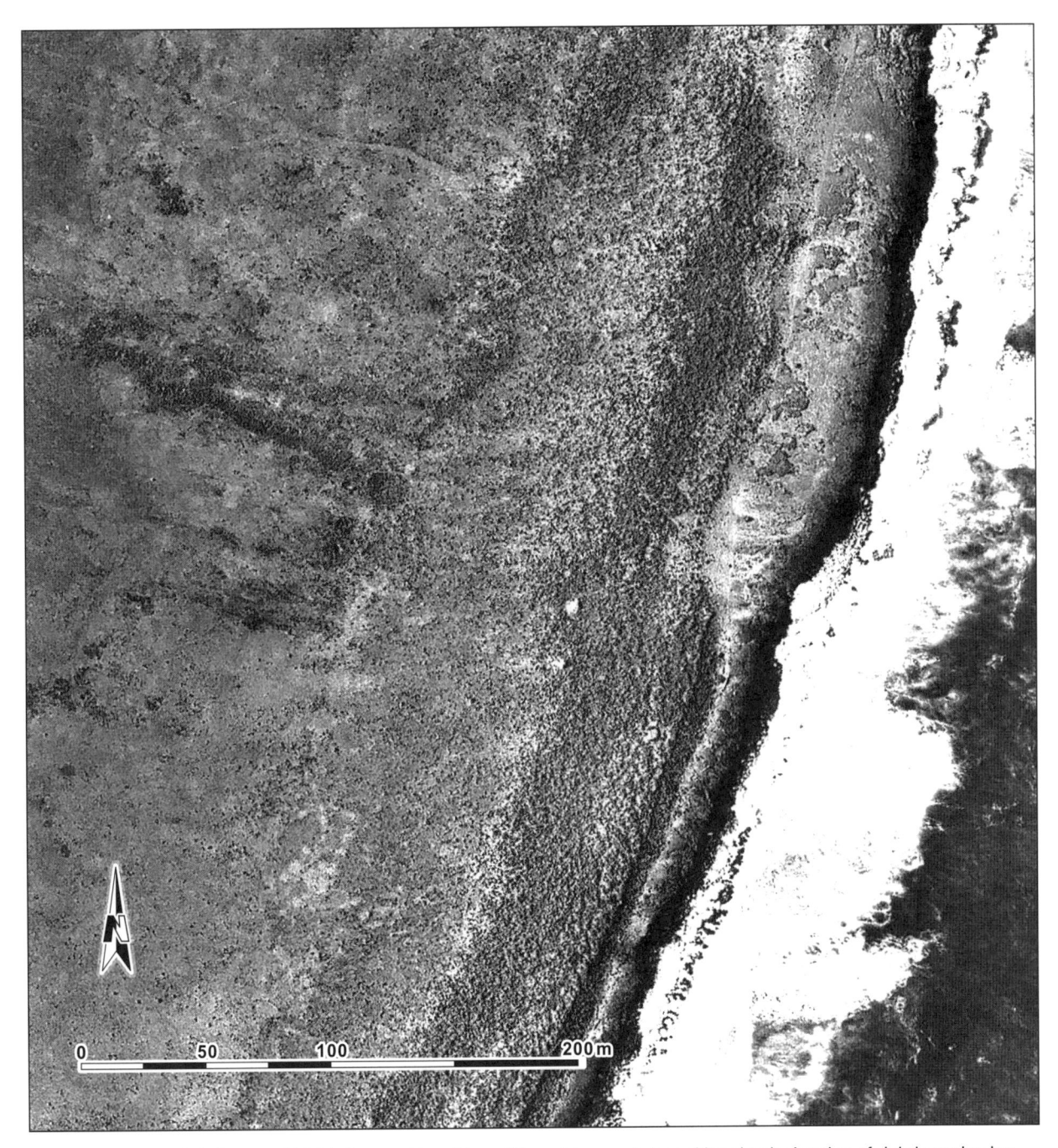

Fig. 116: This aerial picture of 1961 shows at least two different tsunami depositional units by size of debris and color.

Bahamas, see KELLETAT et al., 2005). Another sample from Curaçao has the same radiocarbon age, but the boulders existed in 1634 AD when the Dutch had built their fortification at Piscadera. Tests with *Acropora cervicornis* of the VAN DER HORST coral collection taken in 1920 from Curaçao (SCHEFFERS, 2002a), provided reservoir ages between 260 and 560 years. Our radiocarbon dates have been partly cross-checked with ESR (RADTKE et al., 2002). The difference in the dates was on average 206 years (time span for which ESR dates without any reservoir ages are younger than the radiocarbon data) for nine samples for the period younger than 1,000 years, an average of 293 years for eight data for the period 1000 to 2000 years BP, 286 years for two samples from 2000 to 3000 BP, and 555 years for the period older than 3000 years BP, again only with two samples, not representative so far. More dating is on the way and will be published shortly. All in all we are convinced, that at least for the last 3000 years in the southern Caribbean the reservoir error is much less than 400 years. Since 2002 (SCHEFFERS, 2002a, 2003a, 2004, 2005a; SCHEFFERS et al.,

Fig. 117:
Aerial photograph from 1996 of the mining activities along the coastal stretch of Washikemba.

Fig. 118:
Oblique aerial picture of the Washikemba area with a broad rampart and narrow boulder ridges to inland.

Chapter 4: Tsunami Imprints

Fig. 119: On Curaçao along the leeward side tsunami boulder ridges may show several parallel units of different width, sometimes separated by lines of vegetation.

Fig. 120: Trenches in tsunami boulder ridges at Washikemba and Onima may show at least two units of different age, separated by remnants of a brownish soil in between.

COASTAL RESPONSE TO EXTREME WAVE EVENTS

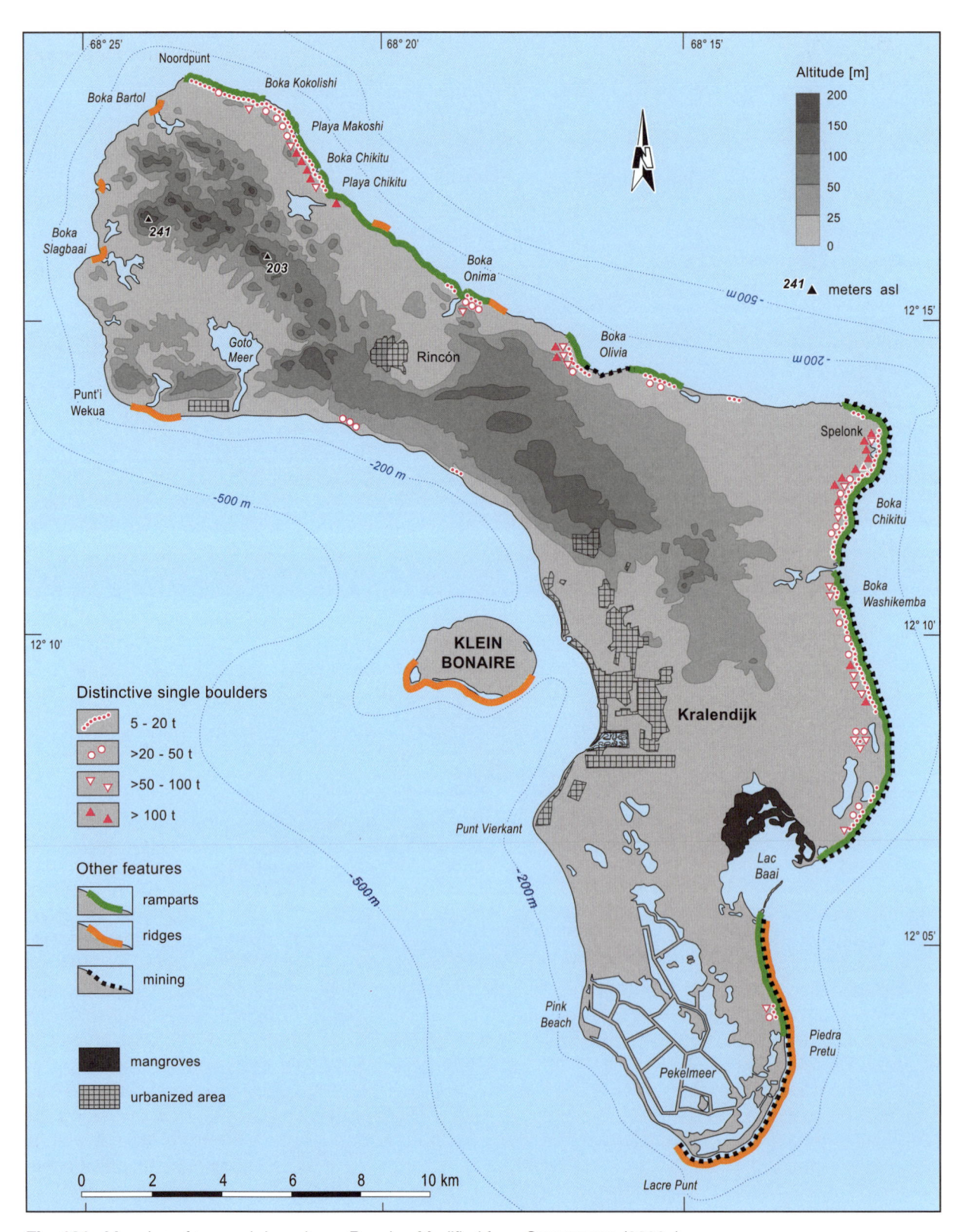

Fig. 121: Mapping of tsunami deposits on Bonaire. Modified from SCHEFFERS (2002a).

Chapter 4: Tsunami Imprints

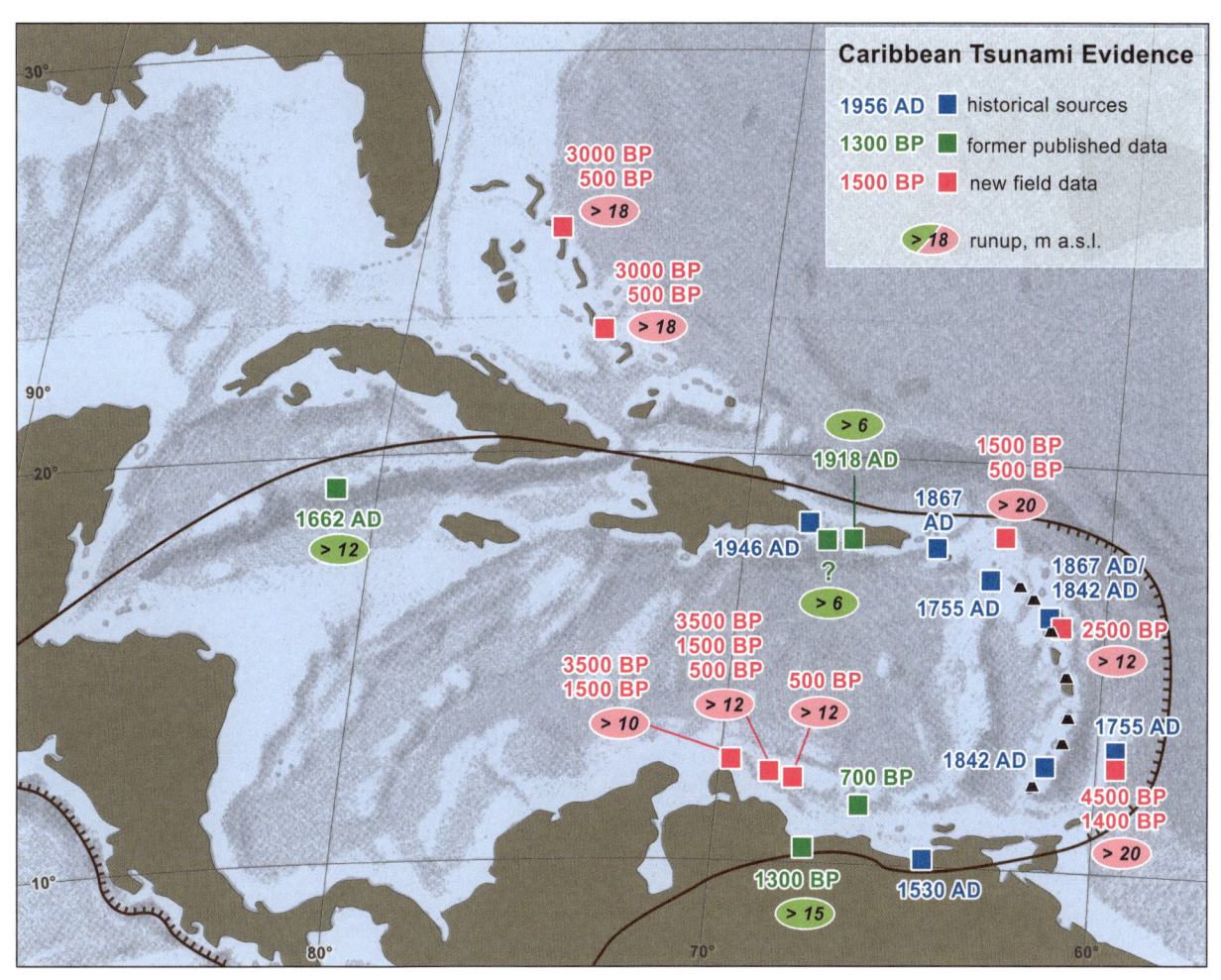

Fig. 122: Historical and absolute tsunami data and minimum run up values for the wider Caribbean.

2005a,c; KELLETAT et al., 2005) about 80 radiocarbon dates and more than 130 ESR dates have been collected from the wider Caribbean, from which more than 120 dates are from the tsunami deposits on Bonaire. They point to very strong tsunami approx. 500 BP, aprox.1200 BP, 3300 BP, 3900 BP and 4300 BP, whereas from Curaçao and Aruba we found dates of 500 BP, 1500 BP, 3500 BP. On Guadeloupe we found dates around 2400 BP, the dates from St. Martin and Anguilla were around 500 BP and 1400 BP, from Barbados 4500 BP and 1400 BP, and from the Bahaman islands of Eleuthera and Long Island around 500 BP and 3000 BP. Combined with the historical dates from the Virgin Island tsunami of 1867, or that of Puerto Rico from 1918, and the published observations from Venezuela, Isla de Mona or Grand Cayman, there is no doubt that a greater number of tsunami with run up values around 15 m and more (Figs. 122-124), inundation at +5-10 m of many 100 m inland, and dislocation of boulders of more than 100 or 200 metric tons are facts for the Intra Americas Seas. Even from these dataset, which is the most extensive one for a certain region worldwide, it is not wise to calculate an average time interval for catastrophic tsunami. But there is no doubt, that strong tsunami can occur every time in all parts of the Caribbean. Establishing a warning system in the Caribbean will not solve the problem definitely – due to the limited warning times – because along the Bahamas and the Antillean island arc (as well as on Aruba, Curaçao and Bonaire) nearly all the tsunami we could identify came from the east, i.e. from the open Atlantic Ocean. A warning system therefore should include possible areas of origin from this region as well.

COASTAL RESPONSE TO EXTREME WAVE EVENTS

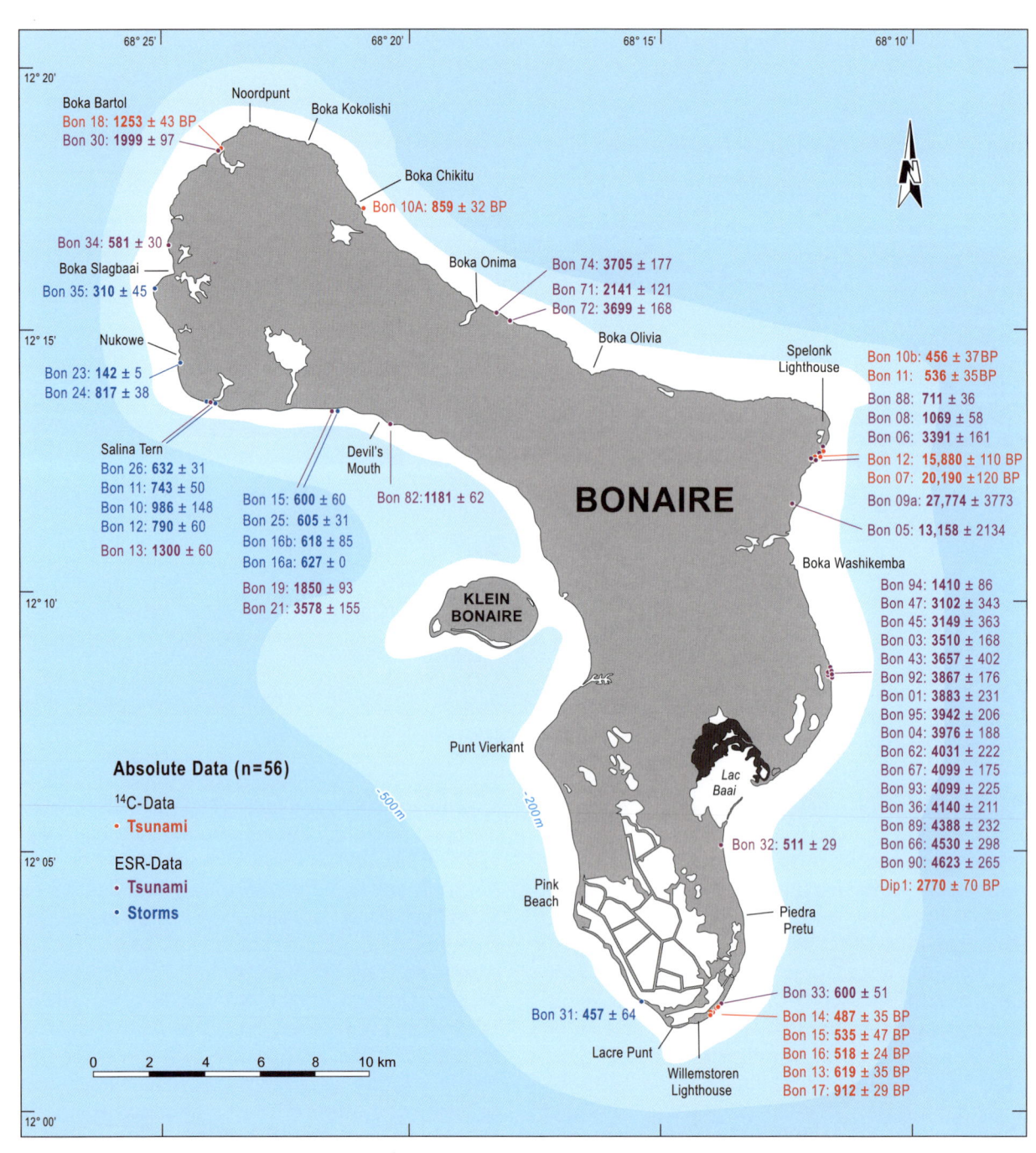

Fig. 123: Absolute data by radiocarbon or ESR for hurricane and tsunami deposits on Bonaire.

Chapter 4: Tsunami Imprints

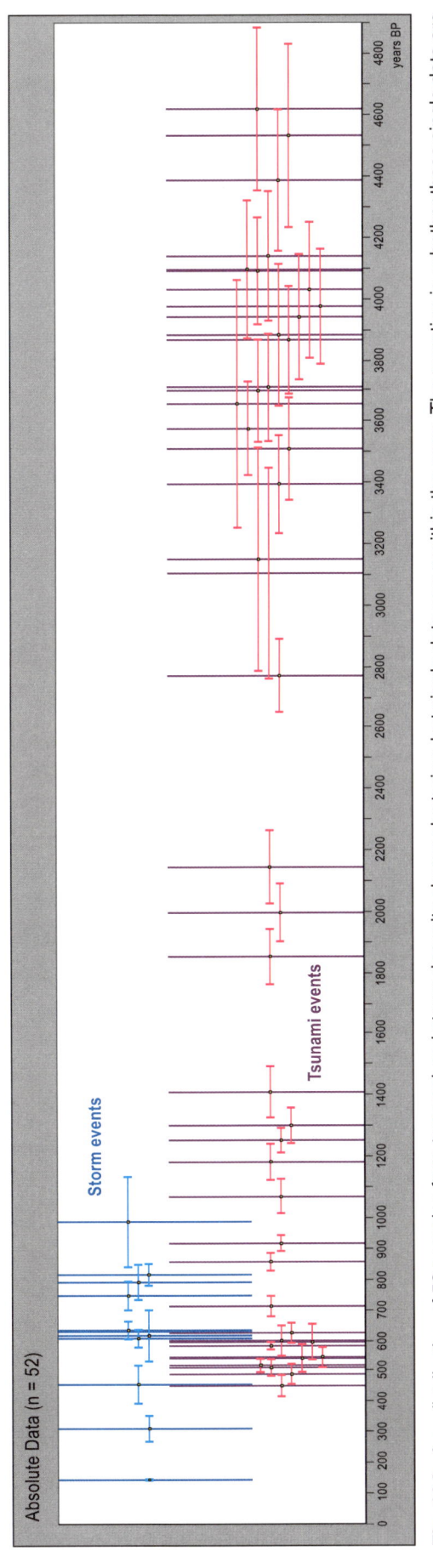

Fig. 124: Age distribution of 52 samples from tsunami and storm deposits shows clustering, but single data occur within the gaps. The question is, whether these single data are representative for events so far not dated significantly, or whether they represent samples of older age buried on the fringing reefs and dislocated with the next tsunami wave.

4.5 Reconstruction of Holocene Reef Communities from Tsunami Deposits

Reconstruction of a Holocene coral reef, in particular in an area with many high magnitude-low frequency events like hurricanes and tsunami, is a difficult task. Normally drilling would give some ideas of the stratigraphy and inner structure, but as the cuts in the uplifted Pleistocene reefs demonstrate (Fig. 125), those drillings may have their difficulties. The species community is variable from place to place in a vertical section: the drilling cores may hit a huge coral like *Montastrea* spec. or a vertical section of *Acropora palmata* and will only show this species for two or more vertical meters, while immediately beside this profile a very complex stratigraphy with rubble, cemented sand and other material may exist. Therefore reconstruction of the Holocene reefs for the dated tsunami events using the debris which has been deposited on land seems a good alternative. Millions of fragments can be investigated, which represent the reef affected at the time of the tsunami impact. We measured profiles across the tsunami ridges or ramparts from the sea to inland, 15 m up to 40 m long, and counted the coarse material in different size classes (= more than 0.5 tons, larger than 30 cm across, smaller than 30 cm across), form type of fragment (= fresh and well preserved, rounded, bored), and species, distinguishing and counting from the well preserved particles *Acropora palmata, Acropora cervicornis* (larger and smaller than 10 cm), *Diploria sp.*, *Montastrea annularis* and *Montastrea cavernosa*.

By this we tested 12 sites in all different exposures on Bonaire, counting approximately 30,000 pieces. We also used this method to distinguish between dated and coarse hurricane deposits and tsunami sediments. The main results will be published soon. Hurricane and tsunami deposits can be distinguished as well as the different exposures. The

Fig. 125:
Cut in uplifted Pleistocene coral reef.

tsunami deposits along the east facing shorelines contain many coarse fragments of *Acropora palmata, Diploria* spec. and *Montastrea* spec., which, on this side of the island, are nearly absent today. Holocene fringing reefs are no longer developed here, but existed till as is documented in the extended tsunami debris. A discussion on the reasons of the disappearing fringing reefs along the trade wind sides including tsunami impacts will be elaborated. It is a fact, that in Holocene times flourishing reefs existed, where nowadays only a bare carbonate platform can be found.

4.6 Comparison of the Bonaire Results with Curaçao and Aruba

SCHEFFERS (2002a,b, 2003a, 2004) has published a lot of field observations and datings from Curaçao and Aruba. The results (concerning amount of debris, run up values, inundation, and amount of coarse debris) are similar to Bonaire with a clear diminishing effect towards Aruba in debris amount and boulder weight. At least three Holocene tsunami have hit the whole area from Bonaire to Aruba, which is a distance of about 200 km, and the run up of these tsunami was greater than +12 m. All along this region tsunami waves have been able to dislocate boulders of more than 50 and 100 tons. Compared with the historical tsunami observed and described, the paleo-tsunami of the Younger Holocene – from 500 BP back to 4300 BP – have had greater power and destructive energy, and it is highly likely, that tsunami of this size will appear again. The source of all these tsunami was in the east, which points to the Antillean Island Arc as a possible origin (plate movements, volcanic explosions, volcanic collapse with submarine slides).

4.7 Comparison of the ABC-Results with the Wider Caribbean

Searching for the origin of the ABC island tsunami we investigated several islands such as Barbados outside the arc, Grenada and St. Lucia on the southern section, Guadeloupe in the center, and St. Martin and Anguilla in the northern part of the Antillean island arc (Fig. 122). Astonishingly, on the volcanic rocks of St. Lucia and Grenada we could not find any large limestone fragments in altitudes and distances to shoreline which would exclude storm impacts, but on both islands Pleistocene tsunami deposits are incorporated into tephra at least up to +50 m asl (SCHEFFERS et al., 2005a). On Barbados, the limestone part of Guadeloupe (i.e. the eastern half), and St. Martin and Anguilla, tsunami boulder deposits are widespread, showing weights of up to 100 tons. On the Bahaman islands of Eleuthera and Long Island the boulders can reach weights of about 300 tons at +15 m of altitude and more than 150 m from the cliff. Their dislocation has been dated to 500 BP and 3000 BP. Again the impressive size of the *Acropora palmata* fragments in these deposits cannot be observed today on the living reefs. Another important observation is, that all the paleo-tsunami identified in the wider Caribbean region evidently came from the east, i.e. from the open Atlantic. The most convincing explanation for this fact may be submarine slides along steeper slopes outside of the Caribbean plate. The mid Atlantic ridge can be excluded as a region for repeated and strong tsunami traveling more than 2,000 km, but volcanic events and submarine slides from the Canary Islands as well as from Madeira or the Azores, could have had these effects. Beside that, the Gorringe Bank SW of Portugal, the source area of the Lisbon event in 1755 AD and another strong tsunami 2400 BP (SCHEFFERS & KELLETAT, 2005) may be responsible for large seaquakes and teletsunami. The slides observed and dated so far on the volcanic island groups, however, show mostly Pleistocene ages, but they have been much stronger than all the Holocene tsunami we could find. Till now not one single imprint made by these huge slides in Pleistocene times had been found.

5. Discussion: The Role of High Magnitude Events on the Coastal Geomorphology and Sedimentology of Bonaire, and Protection Measures

One of the unsolved questions in coastal sciences is: Are gradual processes induced by waves, currents and weathering the most important forming agents, or have single and catastrophic events a greater impact on the forms and sediments along Holocene coasts? In view of our research regarding the Caribbean islands and in particular Bonaire, we can state that single events such as extreme hurricanes and tsunami are the main coastal shaping forces.

The deposits of several Holocene tsunami are well preserved and show very little changes in geomorphological context considering the deposit age (3,000 years and more). Only the finer particles have been partly washed out, and the seaward margin of ramparts and ridges has been cut back by the most extreme hurricane waves. Bokas and other embayments are mostly the result of ingression of the sea into terrestrial forms, combined with breakdown of steep slopes. Along the cliffs, bioerosion in notches and rock pools is much more significant than breakdown by wave impact, which itself can occur during a tsunami, a major storm or every time just by reaching instability along a joint in the coastal rock. The survival time of depositional structures depends on the time span to the next extreme storm or tsunami. The last tsunami appeared several hundred years ago, and the storm ridges and storm terraces in loose rubble of the 1877-hurricane are well preserved, as well as the storm debris of an event about 900 years ago.

Concerning Bonaire, we can conclude that the best protection against tsunami, and especially hurricanes, has been built by tsunami. These are the wide ridges along all the coastlines of Klein Bonaire and along the southern part of Bonaire around Willemstoren and Lacre Punt. The waves of hurricane *IVAN* (2004) did not pass these ridges, only where they have been affected by mining. Therefore these ridges are the best natural protection of Bonaire's coastlines. But we must say that there is no possible protection against run ups comparable with the Holocene paleo-tsunami identified, being +12 m or more, beside a good warning system. If our conclusions on the origin of these tsunami (from the open Atlantic) are right, the warning time would be long enough for almost all the Caribbean islands.

6. Conclusions

Even after several field campaigns, involving investigations on tens of thousands of dislocated fragments, personal observations on hurricane power, and gathering more than 200 absolute dates, the final story of high magnitude-low frequency events on the coastal development of Bonaire and the neighboring islands is not complete. Mining has opened research possibilities to profiles in ridges and ramparts of tsunamigenic origin, yet missing is a full story of the Holocene sedimentation. This can only be gained by analyzing the continuous stratigraphy of sediment traps in the larger embayments, which are open to wave inputs all the time. Another open question is the full story of the Holocene reef development, which should be correlated to the time sections given by Holocene tsunami deposits. Combined, it will give a clear answer to the question why coral reefs are absent along the exposed island sides. And finally, the origin (places and causes) of all the tsunami identified so far is not solved. Only a more extended regional field investigation in the Caribbean and neighboring regions may solve this question.

7. Summary

The Caribbean plate, with 2 cm movement per year and active volcanoes with explosive character on the island arc, is predestined to create strong tsunami in short time intervals. Field observations identified these strong Holocene tsunami in the southern Caribbean, and on Bonaire in particular. The tsunami were dated to the last 4,500 years. The evidence is mostly coarse debris in the form of ridges, ramparts or clusters of large boulders up to 200 tons. Evidence for storm impacts have been identified as well. This material is smaller in size, better sorted and large boulders are missing. These findings are the basis for a better tsunami risk calculation of within the Caribbean and indicate a necessity of a tsunami warning program. The next steps in paleo-tsunami research in the wider Caribbean and elsewhere incorporate the intensification of inductive field research on debatable deposits, and the analysis of drilling profiles within fine sediments containing the total history of the last 6,000 years (last sea level high stand), The findings have to be advertised to the public and should help politicians in deciding how to protect coastal areas from tsunami impacts.

8. Field Trip Itinerary and Facts

The main purpose of the Tsunami Symposium on Bonaire is the presentation of field evidence of storm and tsunami impacts along its coastlines. Therefore we have planned three different trips to cover as many topics as possible. The first two trips are half day tours, the last trip is a full day. During all trips we have to cross unpaved tracks by car, and need stronger shoes for short walks on rough terrain. Figure 126 shows the routes and stops of the field trips.

8.1 First Day, March, 2nd, 2006

Start at Captain Don´s Habitat at 8.00 am for the inspection of the southern coastlines.

1. Stop
West of the Saliñas near White Slave to see the fresh coral rubble ridges built by hurricane LENNY (the Hurricane had a west to east track!) in November, 1999. After more than six years a slight coating by endoliths can be observed. The ridges contain a lot of fresh coral, because this protected side of Bonaire exhibits a living fringing reef dominated by Acropora cervicornis. A short photo stop at White Slave or Red Slave will give the opportunity to see slave huts from colonial times.

2. Stop
Southernmost tip of Bonaire east of Willemstoren lighthouse. Here a well preserved broad tsunami ridge can be seen, mostly built up by coral rubble with 10 to 50 cm in diameter. Reduced by mining from the landward side, the ridge is nearly 50 m wide and more than +2.5 m at the highest parts. In some places, the surface of the deposits shows a micro-topography similar to ripple marks. Seen from the seaside cliff, the deposit shows a lack of sand in the upper parts and the original bimodal texture in the lower parts with very fresh looking gastropods and coral. During hurricane IVAN in September, 2004, the cliff has been abraded by 1-3 m. At the base of the tsunami deposits, the last interglacial coral reef flat is situated near zero meters asl, and is in some places covered by beachrock and coarse ,well rounded, beach particles. The tsunami ridge has been dated to 500- 600 years BP .

3. Stop
Turning north again to visit the coastline with western and southern exposure north of Kralendijk. South of Devil's Mouth we find well preserved notches from the last interglacial highest sea level which are carved in the second to last uplifted reef rock. Stalagmites developed during the last 100.000 years.

The coastline near Stinapa shows tsunami boulders weighing many tons situated on an old reef platform between +3 and +5 m asl. The boulders are interspersed with small Acropora cervicornis fragments. This material has been dislocated by the tsunami or a later storm. At Devil´s Mouth the largest tsunami boulder weighing 70 tons is situated at +6 m and 40 m from the cliff side.

4. Stop
South of Goto Meer. This site shows a tsunami ridge up to +3.5 m with bimodal texture and a huge amount of Acropora cervicornis fragments, partly mixed with boulders of a few tons. A storm layer at +3 m covers these tsunami deposits, showing only coral fragments without sand. Other features present are a steep inactive cliff, and a beach ridge from the storm of 1877 as well as from younger hurricanes. Huge Acropora palmata stems are incorporated in these ridges. A. palmata of this size is not present in the recent coral reef community. Break-off time of these stems is unknown. This section is dated to 600 BP.

5. Stop
Geowatt/Saliña Tern: A steep inactive cliff in uplifted reef rock is accompanied by a tsunami ridge containing huge coral boulders. One of these boulders, a Montastrea cavernosa colony with a diameter of nearly 2 m, has been transported from the living reef at 1300 years BP.

This tsunami ridge is covered by a storm layer up to +3.5 m, dated at 790 years BP and consists of a large amount of Acropora cervicornis. The active cliff has been decorated by a rubble terrace at +2 m from the storm of 1877, and hurricane LENNY of 1999 has left a rubble terrace at about +1 m.

Returning to Captain Don´s Habitat for lunch and the afternoon paper session.

8.2 Second day, March 3rd, 2006

Half day field trip to the east coast of Bonaire.

1. Site with several stops
Driving through the island to south of Washikemba at the east coast. Passing the mining district in tsunami debris, with a lot of large boulders left on site. Main aspect here is the well preserved tsunami ram-

Chapter 8: Field Trip Itinerary and Facts

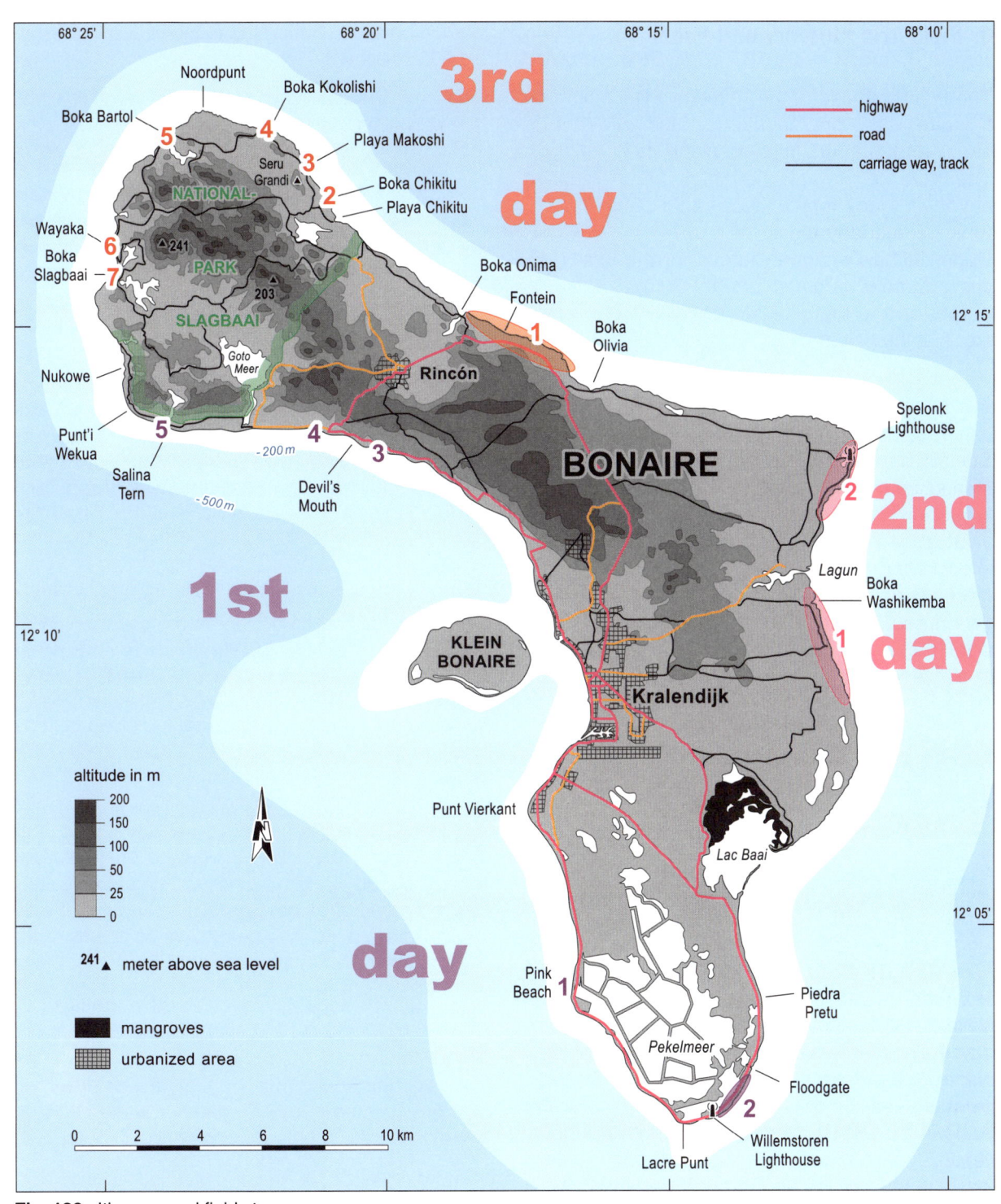

Fig. 126: Itinerary and field stops.

part at the southern end of the mining area. It shows more than 40 m wide debris fields of several kilometers length, including large boulders. The surface of the rampart exhibits ripple marks in coarse debris, best seen on aerial pictures. In this area hurricane *IVAN* of September, 2004, has left some larger boulders from the foreshore, i.e. has moved these boulders for more than 15 m against gravity from 10 m deep water. Other old tsunami boulders have been moved over a short distance. Absolute dates of these deposits – evidently from several tsunami – are 1400 BP, 3100 BP, 3500-4000 BP and 4500 BP.

2. Site with several stops

To the north driving over a lagoonal feature of the last interglacial reef terrace to another rampart area near Spelonk lighthouse. Again much of the deposit has been removed by mining. In the northern section large boulders, many of them weighing more than 100 tons, are situated between 50 and 200 m away from the cliff. Interspersed between these boulders (surrounded by shrubs of *Conocarpus* and other old plants), is coral debris of different size classes. Absolute dating of these events gave 500 BP 3300 BP and Late Pleistocene.

Back to Captain Don´s Habitat for the afternoon paper session.

8.3 Third day, March 4th, 2006

Full day trip to the northeast and Washington-Slagbaai National Park, partly on rough roads, with outside picnic.

1. Sites with several stops
Visiting the tsunami boulders of Olivia Bay, Fontein and Onima Bay, dated to 3700 BP. Hurricane *IVAN* has shifted the front of the ridge for several meters landwards and removed all particles in front of the ridge, leaving a belt of sediment free rock.

The ridges contain many large *Acropora palmata* fragments, and a Holocene reef is missing here. This is in contrast with a very well developed Younger Pleistocene fringing reef, which can be seen with its paleo-cliffs landward of the youngest reef platform.

2. Stop
Boka Chikitu at the northeast coast. In the south the sands of Playa Chikitu can be seen (with a tsunami boulder of 9.20 m length), much of them have been pushed up on the old reef flat by hurricane *IVAN*.

In Boka Chikitu the inner structure of the last interglacial reef can be studied. Landward of this bay, a remnant of an older reef terrace shows a well preserved notch of the last interglacial. Beside this old stack, large boulders have been transported by a Holocene tsunami, the largest one weighing nearly 400 tons. In this Boka a mushroom rock of 22 tons has been broken-off by hurricane *IVAN* in September, 2004.

3. Stop (with picnic)
Seru Grandi: At least three different Pleistocene reefs can be identified here, separated by a good discordance as well as a notch with former coastal caves and younger stalagmites. The area shows a lot of sand moved by hurricane *IVAN* up to at least +8 m asl. This storm partly excavated the old tsunami boulders. Large tsunami boulders and some larger fragments moved by hurricane *IVAN* for 60 m on the reef terrace, as well as fresh cliff-top damage can be seen. Dating the tsunami debris delivered 900 years BP.

4. Stop
Boka Kokolishi. Remarkable in this Boka – beside the exposed inner structure of the Late Pleistocene coral reef – are the beautifully developed young vermetid rims in the inner part which shelter a sandy beach. A large boulder is situated on the reef flat inland of these rims, which evokes the question whether or not these rims have developed after the dislocation of this boulder.

5. Stop
Boka Bartol: a wide ridge separating the Bartol lagoon from the sea. Shown are tsunami deposits dated to 1300 and 2000 years BP, later decorated by a terrace from the 1877 storm and hurricane *LENNY* of 1999.

6. Stop
North of Slagbaai: shown is a spit of nearly 100 m length containing coral debris and constructed by hurricane *LENNY* in 1999, which is somewhat transformed by hurricane *IVAN* in 2004. North of this spit is a semicircular coarse deposit of hurricane *LENNY* on top of a reef terrace at +5.5 m.

7. Stop
Slagbaai: the houses are rebuilt after the destruction by hurricane *LENNY* in 1999. The houses survived every storm since their construction in 1868, but hurricane *LENNY* from the west destroyed this normally protected coastal strip.

Driving back to the southern part of the Nationalpark inland and back to the accommodation for a farewell during sunset.

References

AMERICAN ASSOCIATION OF PETROLEUM GEOLOGISTS (AAPG) (2003): Caribbean Plate Origin. Caribbean Tectonics. –International Meeting, Barcelona Spain, September 21-24.

ADEY, W.H. (1978): Algal ridges of the Caribbean Sea and West Indies. – Phycologia, **17**(4): 361-367.

ADEY, W.H. & BURKE, R.B. (1977): Holocene bioherms of Lesser Antilles – geologic control of development. – Studies in Geology, **4**: 67-81.

ALEXANDER, C.S. (1961): The marine terraces of Aruba, Bonaire and Curaçao, Netherlands Antilles. – Annals of the Association of American Geographers, **51**(1): 102-123.

ALVERSON, K. (2005): Watching over the world's oceans. – Nature, **434**: 19-20.

ATWATER, B.F., TUTTLE, M.P., RAJENDRAN, K. GLAWE, U. & HIGMAN, B. (2005): TSUNAMI GEOLOGY. – Meeting Report, EOS Transactions, American Geophysical Union (in press).

BAK, R.P.M. (1975): Ecological aspects of the distribution of reef corals in the Netherlands Antilles. – Bijdragen tot de Dierkunde, **45**(2): 181-190.

BAK, R.P.M. (1977): Coral reefs and their zonation in Netherlands Antilles. – American Association of Petroleum Geologists, Studies in Geology, Tulsa, Oklahoma.

BANDOIAN, C.A. & MURRAY, R.C. (1974): Pliocene-Pleistocene carbonate rocks of Bonaire, Netherlands Antilles. – Geological Society of America Bulletin, **85**: 1243-1252.

BEETS, D.J. & MAC GILLAVRY, H.J. (1977): Outline of the Cretaceous and Early Tertiary History of Curaçao, Bonaire and Aruba. – Field Guide 8th Caribbean Geological Conference (Curaçao), **15**: 1-6.

BEETS, D.J., MARESCH, W.V., KLAVER, G.T., MONTTANA, A., BOCCHIO, R., BEUNK, F.F. & MONEN, H.P. (1984): Magmatic rock series and high pressure metamorphism as constraints on the tectonic history of the southern Caribbean. – in: BONINI, W.B., HARGARVES, R.B. & SHAGAM, R. (eds.): The Caribbean-South America Plate Boundary and Regional Tectonics. – Geological Society of America Memoirs, **162**: 95-130, Boulder.

BLUME, H. (1974): The Caribbean islands. London (Longman Publishing Group): 464 pp.

BOUDON, G., KOMOROWSKI, J.-C., SEMET, M.-P., LEFRIANT, A. & DEPLUS, C. (1999): Frequent volcanic flank-collapses in the Lesser Antilles Arc: origin and hazards. – American Geographical Union, Fall Meeting 1999, EOS Transactions, AGU 80-46:1142.

BRIES, J.M., DEBROT, A.O. & MEYER, D.L. (2004): Damage to the leeward reefs of Curaçao and Bonaire, Netherlands Antilles from a rare storm event: Hurricane LENNY, November 1999. – Coral Reefs, **23**: 297-307.

BRUCKNER, A.W. & BRUCKNER, R.J. (2003): Condition of Coral Reefs off less Developed Coastlines of Curaçao (Part 1: Stony Corals and Algae). – Atoll Research Bulletin, **496**: 371-393.

BRYANT, E. (2001): Tsunami – the Underrated Hazard. – Cambridge (University Press): 320 pp.

BUSH, D.M. (1991): Impact of Hurricane HUGO on the Rocky Coast of Puerto Rico. – Journal of Coastal Research, **SI, 8**: 49-67.

CARIBBEAN HURRICANE NETWORK (2000): Climatology of Caribbean Hurricanes. – URL: http://stormcarib.com/climatology/freq.htm.

CARRACEDO, J.C. (1999): Growth, structure, instability and collapse of Canarian volcanoes and comparisons with Hawaiian volcanoes. – Journal of Volcanology and Geothermal Research, **94**: 1-19.

CLAGUE J.J., HUTCHINSON I., MATHEWES R.W. & PATTERSON R.T. (1999): Evidence for late Holocene tsunamis at Catala Lake, British Columbia. – Journal of Coastal Research, **15** (1): 45-60.

COCH, N.K. (1994): Geologic effects of hurricanes. – Geomorphology, **10**: 37-63.

CONNELL, J.H. (1997): Disturbance and recovery of coral assemblages. – Coral Reefs, **16**, Suppl.: 101-113.

CRAWFORD, D.A. & MADER, C. (1998): Modeling asteroid impact and tsunami. – Science of Tsunami Hazards, **16**: 21-30.

DAWSON, A.G. (1994): Geomorphological effects of tsunami run-up and backwash. – Geomorphology, **10**: 83-94.

DAWSON, A.G. (1996): The geological significance of tsunamis. – Zeitschrift für Geomorphologie, NF, Suppl. Bd. **102**: 199-210.

DAWSON A.G., LONG D. AND SMITH D.E. (1988): The Storegga Slides: evidence from eastern Scotland for a possible tsunami. – Marine Geology, **82**: 271- 276.

DAWSON, A.G., FOSTER, I.D.L., SHI, S., SMITH, D.E. & LONG, D. (1991): The Identification of Tsunami Deposits in Coastal Sediment Sequences. – Science of Tsunami Hazards, **9**: 73-82.

DAY, S.J., CARRACEDO, J.C., GUILLOU, H. & GRAVESTOCK, P. (1999): Recent structural evolution of the Cumbre Vieja volcano, La Palma, Canary islands: Volcanic rift zone reconfiguration as a precursor to volcanic flank instability. – Journal of Volcanology and Geothermal Research, **94**: 135-167.

DE BUISONJÉ, P.H. (1964): Marine terraces and subaeric sediments on the Netherlands leeward islands Curaçao, Aruba and Bonaire, as indications of Quaternary changes in sea levels and climate. – Koninklijke Nederlandse Academie Wetenschapen, Proc. Ser. **B6**(1): 60-79.

DE BUISONJÉ, P.H. (1974): Neogene and Quaternary Geology of Aruba, Curaçao and Bonaire. – Natuurwetenschappen Studiekring voor Suriname en de Nederlandse Antillen, **78**, Utrecht.

DE BUISONJÉ, P.H. & ZONNEVELD, J.I.S. (1960): De Kustformen von Curaçao, Aruba en Bonaire. – Nieuwe West-Indische Gids, **40**: 121-144.

DE BUISONJÉ, P.H. & ZONNEVELD, J.I.S. (1976): Caracasbaai: A submarine slide of a huge coastal fragment in Curaçao. – Nieuwe West-Indische Grids, **51**: 55-88.

DENGO, G. & CASE, J.E. (eds.) (1990): The Caribbean Region. The geology of North America, Vol. H. – Geological Society of America: 528 pp.

DEPLUS, C., LEFRIANT, A., BOUDON, G., KOMOROWSKI, J.C., VILLEMANT, B., HARFORD, C., SÈGOUFIN, J. & CHEMINÉE, J.-L. (2001): Submarine evidence for large-scale debris avalanches in the Lesser Antilles Arc. – Earth and Planetary Science Letters, **192**: 145-157.

DONE, T.J. & POTTS, D.C. (1992): Influence of habitats and natural disturbances on contributions of massive *Porites* to reef communities. – Marine Biology, **114**: 479-493.

DOLLAR, S.J. & TRIBBLE, G.W. (1993): Recurrent storm disturbance and recovery: A long term study of storm disturbances in Hawaii. – Coral Reefs, **12**: 223-233.

FOCKE, J.W. (1977): The effect of a potentially reef building coralline algal community on an eroding limestone coast, Curaçao (Netherlands Antilles). – Proceedings 3rd International Coral Reef Symposium: 239-245.

FOCKE, J.W. (1978a): Geologische Aspekten van Kwartaire Koraal – en Algenriffen op Curaçao, Bonaire, Aruba en Bermuda. – Ph.D. Thesis, Rijksuniversiteit te Leiden.

FOCKE, J.W. (1978b): Limestone cliff morphology and organism distribution on Curaçao (Netherlands Antilles). – Leidse Geologische Mededelingen, **51**(1): 131-150.

FOCKE, J.W. (1978c): Limestone cliff morphology on Curaçao (Netherlands Antilles), with special attention to the origin of notches and vermetid/coralline algal surf benches („cornices", „trottoirs"). – Zeitschrift für Geomorphologie, NF **22**: 329-349.

FOCKE, J.W. (1978d): Holocene development of coral fringing reefs, leeward off Curaçao and Bonaire (Netherlands Antilles). – Marine Geology, **28**: 31-41.

FOCKE, J.W. (1978e): Subsea (0-40 m) terraces and benches, windward of Curaçao, Netherlands Antilles. – Leidse Geologische Mededelingen, **51**(1): 95-102.

FOSTER, I.D.L., ALBON, A.J., BARDELL, K.M., FLETCHER, J.L., JARDINE, T.C., MOTHERS, R.J., PRITCHARD, M.A. & TURNER, S.E. (1991): High energy coastal sedimentary deposits; an evaluation of depositional processes in Southwest England. – Earth Surface Processes Landforms, **16** (4): 341-356.

FOUKE, B. (1994): Deposition, Diagenesis and Dolomitization of Neogene Seroe Domi Formation Coral Reef Limestones on Curaçao, Netherlands Antilles. – Uitgaven Natuurwetenschappelijke Studiekring voor het Caraibisch Gebied, **134**: 182 pp.

GARDNER, T., CÔTÉ, I.M., GILL, J.A., GRANT, A., WATKINSON, A.R. (2005): Hurricanes and Caribbean coral reefs, impacts, recovery patterns, and role in long-term decline. – Ecology, **86**: 174-184.

GEISTER, J. (1980): Calm-water reefs and rough-water reefs of the Caribbean Pleistocene. – Acta Paleontologica Polonica, **25**: 541-545.

GEISTER, J. (1983): Holozäne westindische Korallenriffe: Geomorphologie, Ökologie und Fazies. – Facies, **9**: 173-284.

GOFF, J.R., CROZIER, M., SUTHERLAND, V., COCHRAN, U. & SHANE, P. (1998): Possible tsunami deposits from the 1855 earthquake, North Island. – In: STEWART, I.S & VITA-FINZI, C. (eds): Coastal Tectonics. – Geological Society, London, Special Publications, **146**: 353-374.

GOFF, J., CHAGUÉ-GOFF, C. & NICHOL, S. (2001): Paleotsunami deposits: a New Zealand perspective. – Sedimentary Geology, **143**: 1-6.

GOFF, J.R., MCFADGEN, B.G., CHAGUE-GOFF, C. (2004): Sedimentary differences between the 2002 Easter storm and the 15th century Okoropunga tsunami, Southeastern North Island, New Zealand. – Marine Geology, **204**: 235-250.

GOREAU, T.F. & LAND, L.S. (1974): Fore-reef morphology and depositional processes, north Jamaica. – In: Laporte, L.F. (ed.): Reefs in Time and Space. – Society Economic Paleontologists and Mineralogists, Tulsa, Oklahoma: 77-89.

GUINEY, J.L. (2000): Preliminary Report: Hurricane *LENNY*, 13-23 November 1999. – National Hurricane Center, Miami, Florida, URL: http:// www.nhc.noaa.gov/ 1999Lenny.html.

HARTOG, J. (1968): Curaçao, from colonial dependance to autonomy. – History of the Netherlands Antilles, **3**.

HAYNE, M. & CHAPPELL, J. (2001): Cyclone frequency during the last 5,000 yrs from Curacoa Island, Queensland. – Palaeogeography, Palaeoclimatology, Palaeoecology, **168**: 201-219.

HEARTY, J.P. (1997): Boulder deposits from large waves during the last interglaciation on North Eleuthera island, Bahamas. – Quaternary Research, **48**: 326-338

HEINRICH, P., GUIBOURG, S., MANGENEY, A. & ROCHE, R. (1999): Numerical modeling of a landslide-generated tsunami following a potential explosion of the Montserrat Volcano. – Physics and Chemistry of the Earth, (A), **24**(2): 163-168.

HEMPHILL-HALEY, E. (1996): Diatoms as an aid in identifying late-Holocene tsunami deposits. – Holocene, **6** (4): 439-448.

HERWEIJER, J.P. & FOCKE, J.W. (1978): Late Pleistocene Depositional and Denudational History of Aruba, Bonaire and Curaçao (Netherlands Antilles). – Geologie en Mijnbouw, **57**: 177-187.

HOLLAND, G.J. (1997): The Maximum Potential Intensity of Tropical Cyclones. – Journal of the Atmospheric Sciences, **54**: 2519-2541.

HUBBARD, D.K., SADD, J.L. & ROBERTS. H.H. (1981): The role of physical processes in controlling sediment transport patterns on the insular shelf of St. Croix, U.S. Virgin Islands. – Proceedings of the Fourth International Coral Reef Symposium, 399-404.

HUBBARD, D.K., PARSONS, K.M., BYTHELL, J.C. & WALKER, N.D. (1991): The Effects of Hurricane *HUGO* on the Reefs and Associated Environments of St. Croix, U.S. Virgin Islands – A Preliminary Assessment. – Journal of Coastal Research, **S1** (8):33-48.

JONES, B. & HUNTER, G. (1992): Very large Boulders on the Coast of Grand Cayman: The Effects of Giant waves on Rocky Shorelines. – Journal of Coastal Research, **8**: 763-774.

KELLETAT, D. (1997): Mediterranean coastal biogeomorphology: processes, forms and sea-level indicators. – *in*: BRIAND, F. & MALDONADO, A. (eds.), Transformations and evolution of the Mediterranean coastline. – CIESM Science Series, No.3, Bulletin de l´Institut Océanographique, Monaco, Numéro Special **18**: 209-226.

KELLETAT, D. (2003): Tsunami durch Impact von Meteoriten im Quartär? – Essener Geographische Arbeiten, **35**: 27-38.

KELLETAT, D. & SCHEFFERS, A. (2001): Hurricanes und Tsunamis - Dynamik und küstengestaltende Wirkungen. – Bamberger Geographische Schriften, **20**: 29-53.

KELLETAT, D. & SCHELLMANN, G. (2001): Sedimentologische und geomorphologische Belege starker Tsunami-Ereignisse jung-historischer Zeitstellung im Westen und Südosten Zyperns. – Essener Geographische Arbeiten, **32**: 1-74.

KELLETAT, D. & SCHELLMANN, G. (2002): Tsunamis on Cyprus – Field Evidences and ^{14}C Dating Results. – Zeitschrift für Geomorphologie, NF, **46**(1): 19-34.

KELLETAT, D. & SCHEFFERS, A. (2004a): Sedimentologische und geomorphologische Tsunamispuren an den Küsten der Erde. – HGG-Journal, Jahrbuch Heidelberger Geographische Gesellschaft, **18**: 11-20.

KELLETAT, D. & SCHEFFERS, A. (2004b): Tsunami im Atlantischen Ozean? – Geographische Rundschau, **56**(6): 4-12.

KELLETAT, D. & SCHEFFERS, A. (2004c): Bimodal tsunami deposits – a neglected feature in paleo-tsunami research. – *in*: SCHERNEWSKI, G. & DOLCH, T: Geographie der Meere und Küsten. – Coastline Reports, **1**: 1-20.

KELLETAT, D., SCHEFFERS, A. & SCHEFFERS, S. (2005): Holocene tsunami deposits on the Bahaman islands of Long Island and Eleuthera. – Zeitschrift f. Geomorphologie, NF **48**(4): 519-540.

KLAVER, G.T. (1987): The Curaçao Lava Formation: an ophiolithic analogue of the anomalous thick layer 2B of the Mid-Cretaceous oceanic plateaus in the Western Pacific and Central Caribbean. – GUA Papers of Geology, **27**: 1-68.

KOBLUK, D.R. & LYSENKO, M.A. (1984): Carbonate rocks and coral reefs Bonaire, Netherlands Antilles Field Trip Guidebook 13. – Geological Association of Canada – Mineralogical Association of Canada, Ottawa.

KOBLUK, D.R. & LYSENKO, M.A. (1992): Storm features on a southernCaribbean fringing coral reef. – Palaios, **7**: 213-221.

KORTEKAAS S. (2002): Tsunami, storms and earthquakes: Distinguishing coastal flooding events. – PhD Thesis, University of Coventry, Coventry, 228p.

KRASTEL, S., SCHMINCKE, H.U., JACOBS, C.L., RIHM, R., LE BAS, T.P. & ALIBES, B. (2001): Submarine landslides around the Canary islands. – Journal of Geophysical Research, Solid Earth, **106** (B3): 3977-3997.

LANDER, J. & WHITESIDE, L.S. (1997): Caribbean Tsunami, an Initial History. – Mayaguez Tsunami Workshop, June 11-13, 1997, Puerto Rico.

LANDER, J.F., WHITESIDE, L.S. & LOCKRIDGE, P.A. (2002): A brief history of tsunamis in the Caribbean. – Science of Tsunami Hazards, **20**(2): 57-94.

LEATHERMAN, S.P. & WILLIAMS, A.T. (1977): Lateral textural grading in washover sediments. – Earth Surface Process and Landforms, **2**: 333-341.

LIGHTY, R.G., MACINTYRE, I.G. & STUCKENRATH, R. (1982): *Acropora palmata* reef framework: A reliable indicator of sea level in the Western Atlantic for the past 10,000 years. – Coral Reefs, **1**: 125-130.

LUDLUM, D.M. (1989): Early American Hurricanes, 1492-1870. – Pennsylvania (Lancaster Press): 198 pp.

MACDONALD, R., HAWKESWORTH, C.J. & HEATH, E. (2000): The Lesser Antilles Volcanic Chain: a study in arc magmatism. – Earth Science Reviews, **69**: 1-79.

MANN, P., SCHUBERT, C. & BURKE, K. (1990): Review of Caribbean Neotectonics. – *in*: DENGO, G. & CASE, J.E. (eds.): The Caribbean Region. The Geology of North America, Vol. 2. – Geological Society of America, Boulder: 207-238

MASTRONUZZI, G. & SANSO, P. (2000): Boulder transport by catastrophic waves along the Ionian coast of Apulia, Southern Italy. – Marine Geology, **170**: 93-103.

METEOROLOGICAL SERVICE OF THE NETHERLANDS ANTILLES AND ARUBA (MSNAA) (2002): Hurricanes and tropical storms in the Netherlands Antilles and Aruba. – URL: http://www.meteo.an/meteo2/eng/reports/hurtrop1.htm.

MEYER, D.L., BRIES, J.M., GREENSTEIN, B.J. & DEBROT, A.O. (2003): Preservation of in situ framework in regions of low hurricane frequency: Pleistocene of Curaçao and Bonaire, southern Caribbean. – Lethaia, 36: 273-286.

MILLAS, J.C. (1968): Hurricanes of the Caribbean and adjacent regions, 1492-1800. – Academy of Arts and Sciences of the Americas, Miami, 328 pp.

MINOURA, K. & NAKATA, T. (1994): Discovery of an Ancient Tsunami Deposit in Coastal Sequences of Southwest Japan: Verification of a Large Historic Tsunami. – Island Arc, 3: 66-72.

MINOURA, K., NAKAYA, S. & UCHIDA, M. (1994): Tsunami deposits in a lacustrine sequence of the Sanriku coast, northeast Japan. – Sedimentary Geology, 89 (1-2): 25-31.

MOREIRA, V.S. (1993): Historical tsunamis in mainland Portugal and Azores – Case histories. – In Tinti, S. (ed) Tsunamis in the World. – 15th International Tsunami Symposium, 1991: 65-73.

MOORE, G.W. & MOORE, J.G. (1988): Large scale bedforms in bolder gravel produced by giant waves in Hawaii. –GSA Special Paper, 229: 101–109.

MOORE J.G., NORMARK W.R. & HOLCOMB R.T. (1994): Giant Hawaiian Landslides. – Annual Reviews Earth Planetary Science, 22: 119-144.

MOYA, J.C. (1999): Stratigraphical and morphologic evidence of tsunami in northwestern Puerto Rico. – Sea Grant College Program, University of Puerto Rico, Mayaguez Campus.

NANAYAMA, F., SHIGENO, K., SATAKE, K., SHIMOKAWA, K., KOITABASHI, S., MAYASAKA, S. & ISHII, M. (2000): Sedimentary differences between 1993 Hokkaido-Nansei-Oki tsunami and 1959 Miyakijima typhoon at Tasai, southwestern Hokkaido, northern Japan. – Sedimentary Geology, 135: 255-264.

NEUMANN, A.C. & MACINTIRE, I. (1985): Reef response to sea level rise: keep-up, catch-up or give-up. – Proceedings 5th International Coral Reef Congress, Tahiti, Vol. 3: 105-110.

NEUMANN, C.J., JARVINEN, B.R., MCADIE, C.J. & ELMS, J.D. (1993): Tropical cyclones of the North Atlantic Ocean, 1871-1992. – National Climatic Data Center in cooperation with the National Hurricane Center, Coral Gables, Florida, 193 pp.

NISHIMURA, Y. & MIYAJI, N. (1995): Tsunami Deposits from the 1993 Southwest Hokkaido Earthquake and the 1640 Hokkaido Komagatake Eruption, Northern Japan. – Pageoph, 144(3/4): 719-733.

NOAA (2004): URL: http://.hurricane.csc.noaa.gov/hurricanes/hurrPrint.htm.

NOTT, J. (1997): Extremely high-energy wave deposits inside the Great Barrier Reef, Australia: determining the cause – tsunami or tropical cyclone. – Marine Geology, 141: 193-207.

NOTT, J. (2000): Records of Prehistoric Tsunamis from Boulder Deposits – Evidence from Australia. – Science of Tsunami Hazards, 18(1): 3-14.

NOTT, J. (2003a): Waves, coastal boulders and the importance of the pre-transport setting. – Earth and Planetary Science Letters, 210: 269-276.

NOTT, J. (2003b): Tsunami or storm waves? – determining the origin of a spectacular field of wave emplaced boulders using numerical storm, surge and wave models and hydrodynamic transport equations. – Journal of Coastal Research, 19: 348-356.

NOTT, J. (2004): The tsunami hypothesis – comparison of field evidence against the effects, on the Western Australian coasts, of some of the most powerful storms on Earth. – Marine Geology, 208: 1-12.

NOTT, J. & BRYANT, E. (2003): Extreme marine inundations (tsunamis ?) of coastal Western Australia. – Journal of Geology, 111: 691-706.

NOTT, J. & HAYNE, M. (2001): High frequency of „super-cyclones" along the Great Barrier Reef over the past 5,000 years. – Nature, 413: 508-512.

OTA, Y., PIRAZZOLI, P.A., KAWANA, T. & MORIWAKI, H. (1985): Late Holocene coastal morphology and sea-level records on three small islands, the South Ryukyus, Japan. – Geographical Review of Japan, Series B 58, 2: 185-194.

PANDOLFI, J.M., LLEWELLYN, G. & JACKSON, J.B.C. (1999): Pleistocene reef environments, constituent grains, and coral community structure: Curaçao, Netherlands Antilles. – Coral Reefs, 18: 107-122.

PANDOLFI, J.M. & JACKSON, J.B.C. (2001): Community structure of Pleistocene coral reefs of Curaçao, Netherlands Antilles. – Ecological Monographs, 71: 49-67.

PARARAS-CARAYANNIS, G. (2004): Volcanic Tsunami Generation Source Mechanisms in the Eastern Caribbean Region. – Science of Tsunami Hazards, 22: 74-114.

PATTERSON, R.T. & FOWLER, A.D. (1996): Evidence of self organization in planktic foraminiferal evolution: implications for interconnectedness of paleoecosystems. – Geology 24 (3): 215-218.

SEDGEWICK, P.E. & DAVIS, R.A. (2003): Stratigraphy of washover effects in Florida: implications for recognition in the stratigraphic record. – Marine Geology, 200: 31-48.

PINDALL, J.L. & BARRETT, S.F. (1990): Geological evolution of the Caribbean region: a plate tectonic perspective. – in: DENGO, G. & CASE, J.E. (eds.): The Geology of North America: 405-432.

RADTKE, U., SCHELLMANN, G., SCHEFFERS, A., KELLETAT, D., KROMER, B. & KASPER, H.U. (2002): Electron spin resonance and radiocarbon dating of coral deposited by Holocene tsunami events on Curaçao, Bonaire and Aruba (Netherlands Antilles). – Quaternary Science Reviews, 22: 1305-1317.

RAPPAPORT, E.N. & FERNANDEZ-PARTAGAS, J.F. (1997): The deadliest Atlantic tropical cyclones, 1492-present. – URL: http://nhc.noaa.gov/pastdeadlya1.html.

READING, A.J. (1990): Caribbean tropical storm activity over the past few centuries. – International Journal of Climatology, **10**: 365-376.

ROBERTS, H.H. (1974): Variability of reefs with regard to changes in wave power around an island: Brisbane, Australia. – Proceedings of the Second International Coral Reef Symposium, **2**: 497–512.

ROBERTS, H.H. (1989): Physical processes as agents of sediment transport in carbonate systems: examples from St. Croix, USVI. – In: Hubbard, D.K. (ed.): Terrestrial and Marine Geology of St. Croix, U.S. Virgin Islands. –St. Croix, USVI, WIL Special Publication, **8**, West Indies Laboratory.

ROBSON, G.R. (1964): An earthquake catalogue for the Eastern Caribbean, 1530-1960. – Bulletin of the Seismological Society of America, **54**: 785-832.

ROOBOL, M.J., WRIGHT, J.V. & SMITH, A.L. (1983): Calderas or gravity-slides structures in the Lesser Antilles Island Arc? – Journal of Volcanology and Geothermal Research, **19**: 121-134.

ROOS, P.J. (1971): The shallow-water stony corals of the Netherlands Antilles. Studies on the Fauna of Curaçao and other Caribbean Islands. – Natuurwetenschappen Studiekring voor Suriname en de Nederlandse Antillen, **130**, Utrecht.

RULL, V. (2000): Holocene sea level rising in Venezuela: a preliminary curve. – Boletim Sociedad Venezolana Geologos: URL: http://www.ecopal.org/sealevel.htm

SATO, H., SHIMAMOTO, T., TSUTSUMI, A. & KAWAMOTO, E. (1995): Onshore Tsunami Deposits Caused by the 1993 Southwest Hokkaido and 1993 Japan Sea Earthquakes. – Pageoph, **144**(3/4): 693-717.

SCATTERDAY, J.W. (1974): Reefs and associated coral assemblages off Bonaire, Netherlands Antilles, and their bearing on Pleistocene and recent reef models. – Proceedings 2nd International Coral Reef Symposium, Brisbane, October 1974.

SCHEFFERS, A. (2002a): Paleosunami in the Caribbean: Field Evidences and Datings from Aruba, Curaçao and Bonaire. – Essener Geographische Arbeiten, **33**.

SCHEFFERS, A. (2002b): Paleotsunami Evidences from Boulder Deposits on Aruba, Curaçao and Bonaire. – Science of Tsunami Hazards, **20**(1): 26-37.

SCHEFFERS, A. (2003a): Landschaftsspuren und Zeitstellung holozäner Tsunami auf den Niederländischen Antillen (Aruba, Curaçao und Bonaire). – Berichte Forschungs- und Technologiezentrum Westküste, Büsum, **28**: 75-92

SCHEFFERS, A. (2003b): Boulders on the Move: Beobachtungen aus der Karibik und dem westlichen Mittelmeergebiet. – Essener Geographische Arbeiten, **35**: 2-10.

SCHEFFERS, A. (2004): Tsunami imprints on the Leeward Netherlands Antilles (Aruba, Curaçao and Bonaire) and their relation to other coastal problems. – Quaternary International, **120**(1): 163-172.

SCHEFFERS, A. (2005a): Argumente und Methoden zur Unterscheidung von Sturm- und Tsunamischutt und das Problem der Datierung von Paläo-Tsunami. – Die Erde (in press).

SCHEFFERS, A. (2005b): Windskulpturierung (ventifaction) im Küstenraum von West-Portugal und am SE-Peloponnes, Griechenland. – Schriften des Arbeitskreises Landes- und Volkskunde, Koblenz (in press).

SCHEFFERS, A. & KELLETAT, D. (2003): Sedimentologic and Geomorphologic Tsunami Imprints Worldwide – A Review. – Earth Science Reviews, **63**(1-2): 83-92.

SCHEFFERS, A. & KELLETAT, D. (2005): Tsunami Relics in the Coastal Landscape West of Lisbon, Portugal. – Science of Tsunami Hazards, **23**(1): 3-16.

SCHEFFERS, A., SCHEFFERS, S. & KELLETAT, D. (2005a): Paleo-Tsunami Relics on the Southern and Central Antillean Island Arc (Grenada, St. Lucia and Guadeloupe). – Journal of Coastal Research, **21**(2): 263-273.

SCHEFFERS, A., SCHEFFERS, S. & KELLETAT, D. (2005b): Documentation of the Impact of Hurricane IVAN on the Coastline of Bonaire (Netherlands Antilles). – Journal of Coastal Research (submitted).

SCHEFFERS, S. (2005): Benthic-pelagic coupling in coral reefs: Interaction between Framework Cavities and Reef Water. – PhD thesis, University of Amsterdam, Göttingen (Shaker Verlag): 114 pp.

SCHELLMANN, G., RADTKE, U. & WHELAN, F. (2004): The Marine Quaternary of Barbados. – Kölner Geographische Arbeiten, **81**: 137 pp.

SCHELLMANN, G., RADTKE, U., SCHEFFERS, A., WHELAN, F. & KELLETAT, D. (2004): ESR dating of coral reef terraces on Curaçao (Netherlands Antilles) with estimates of Younger Pleistocene sea level elevations. – Journal of Coastal Research, **20**(4): 947-957.

SCHIERECK, G.J., BOOIJ, N. & HOLTHUIJSEN, L.H. (1997): Water Movement During a Hurricane Near Bonaire. – Delft University of Technology, The Netherlands, 11pp.

SCHUBERT, C. (1988): Neotectonics of the La Victoria Fault Zone, North-Central Venezuela. – Annales Tectonicae, **II**: 58-66.

SCHUBERT, C. (1994): Tsunamis in Venezuela; Some Observations on Their Occurrence. – Journal of Coastal Research, Spec. Issue, **12**: 189-195.

SCOFFIN, T.P. (1993): The geological effects of hurricanes on coral reefs and the interpretation of storm deposits. – Coral Reefs, **12**: 203-221.

SIGURDSSON, H. & CAREY, S. (1991): Caribbean Volcanoes: A Field Guide. – Geological Association of Canada: 101 pp.

SHENNAN, I., RUTHERFORD, M.M., INNES, J.B. & WALKER, K.J. (1996): Late glacial sea level and ocean margin environmental changes interpreted from biostratigraphic and lithostratigraphic studies of isolation basins in northwest Scotland. – Geological Society, London; Special Publication, **111**: 229-244.

SHI, S., DAWSON, A.G. & SMITH, D.E. (1995): Coastal Sedimentation Associated with the December 12th, 1992 Tsunami in Flores, Indonesia. – Pageoph, **144** (3/4): 525-536.

SHINN, E.A. (1963): Spur and groove formation on the Florida reef tract. – Journal of Sedimentary Petrology, **33**: 291-303.

SMITH, M.S. & SHEPHERD, J.B. (1993): Preliminary investigations of the tsunami hazard of Kick´em Jenny submarine volcano. – Natural Hazards, **7**: 257-277.

SOROKIN, Y.I. (1993): Coral reef ecology. – New York (Springer-Verlag).

SPENCER, T. & VILES, H. (2002): Bioconstruction, bioerosion and disturbance on tropical coasts: coral reefs and rocky limestone shores. – Geomorphology, **48**: 23-50.

STODDART, D.R. (1962): Catastrophic storm effects on the British Honduras reefs and cays. – NATURE **196**: 512-515.

STODDART, D.R. (1971): Coral reefs and islands and catastrophic storms. – in: STEERS, J.A. (ed.): Applied Coastal Geomorphology, Cambridge, Massachusetts (MIT Press): 155-197.

STODDART, D.R. (1974): Post hurricane changes on the Britsih Honduras reefs: resurvey of 1972. – Proceedings 2nd International Coral Reef Symposium, Brisbane, **2**: 473-483.

STORMCARIB (2005): Climatology of Caribbean Hurricanes. – URL: http://stormcarib.com/climatology/TNCB_all_isl.htm.

STUIVER, M., REIMER, P. & BRAZIUNAS, T.F. (1998): High-precision radiocarbon age calibration for terrestrial and marine samples. – Radiocarbon, **40**: 1127-1151.

TAGGART, B.E., LUNDBERG, J.L., CAREW, L. & MYLROIE, J.E. (1993): Holocene reef-rock boulders on Isla de Mona, Puerto Rico, transported by a hurricane or seismic sea wave. – Geological Society of America, Abstract with Programs **25**(H6): A-61.

TOMBLIN, J.F. (1975): The Lesser Antilles and Aves Ridge. – in: Nairn, A.E.M. & Stehli, F.G. (eds.): The Ocean Basins and Margins, Vol. 3: Gulf of Mexico and the Caribbean. – New York (Plenum Press): 467-500.

TUTTLE, M.P., RUFFMAN, A., ANDERSON, T. & JETER, H. (2004): Distinguishing tsunami from storm deposits in eastern North America: the 1929 Grand Banks tsunami versus the 1991 Halloween storm. – Seismological Research Letters, **75**: 117-131.

UNISYS CORPORATION (2002): Atlantic tropical storm tracking by year. – URL: http://weather.unisys.com/hurricane/atlantic/index.html.

VAN DUYL, F. (1985): Atlas of the living reefs of Curaçao and Bonaire (Netherlands Antilles). – Natuurwetenschappen Studiekring voor Suriname en de Nederlandse Antillen, **117**, Utrecht.

VAN LOENHOUD, P.J. & VAN DE SANDE, J.C.P.M. (1977): Rocky Shore Zonation in Aruba and Curaçao (Netherlands Antilles), with the Introduction of a new General Scheme of Zonation. Vol. I. – Proceedings Koninklijke Nederlandse Akademie Wetenschappen, Series C – Biological and Medical Sciences, **80**(1): 437-474.

VAN'T HOF, T. (1997): New Guide to the Bonaire Marine Park. – Harbour Village Beach Resort, Bonaire, 207 pp.

VAN VEGHEL, M.L.J. (1996): A field guide to the reefs of Curaçao. – 8th International Coral Reef Symposium, 24th June 1996, Panama City.

VAN VEGHEL, M.L.J. & HOETJES, P.C. (1995): Effects of Tropical Storm *BRET* on Curaçao Reefs. – Bulletin Marine Sciences, **56**: 692-694.

WARD, ST. & ASPHAUG, E. (2000): Asteroid Impact Tsunami: A Probabilistic Hazard Assessment. – Icarus, **145**: 64-78.

WEISS, M.P. (1979): A saline lagoon on Cayo Sal, western Venezuela. - Atoll Research Bulletin, **232**: 1-33.

WEYL., R. (1966): Geologie der Antillen. – Berlin (Borntraeger): 410 pp.

WHELAN, F. & KELLETAT, D. (2002): Geomorphic Evidence and Relative and Absolute Dating Results for Tsunami Events on Cyprus. – Science of Tsunami Hazards, **20**(1): 3-18.

WHELAN, F. & KELLETAT, D. (2003a): Analysis of Tsunami Deposits at Cabo de Trafalgar, Spain, Using GIS and GPS Technology. – Essener Geographische Arbeiten, **35**: 11-25.

WHELAN, F. & KELLETAT, D. (2003b): Submarine Slides on Volcanic Islands – A source for Mega-Tsunamis in the Quaternary. – Progress in Physical Geography, **27**(2): 198-216.

WOOD, R. (2001): Biodiversity and the history of reefs. – Geological. Journal, **36**: 251-263.

ZAHIBO, N. & PELINOVSKY, E.N. (2001): Evaluation of tsunami risks in the Lesser Antilles. – Natural Hazards and Earth System Sciences, **1**: 221-231.

List of Figures

Fig. 1: Geodynamics of the Caribbean (SCHEFFERS, 2002a, modified from SCHUBERT, 1988; MANN et al., 1990; and others).

Fig. 2: *Millepora* spec.

Fig. 3: *Porites porites*

Fig. 4: *Madracis mirabilis*

Fig. 5: *Montastrea annularis*

Fig. 6: (a) *Diploria labyrinthiformis*, (b) *Diploria strigosa*

Fig. 7: *Acropora palmata*

Fig. 8: *Montastrea cavernosa*

Fig. 9: *Colpophylla natans*

Fig. 10: *Acropora cervicornis*

Fig. 11: *Agaricia* spec.

Fig. 12: Storm tracks in the Intra Americas Seas from 1886 to 1995 (METEOROLOGICAL SERVICE OF THE NETHERLANDS ANTILLES AND ARUBA, 2002).

Fig. 13: Tsunami in the Caribbean since 1530 (modified after LANDER, WHITESIDE & LOCKRIDGE, 2002). The map shows only tsunami events, which are considered probable (V3) or reliable (V4) by the authors.

Fig. 14: Geological map of Bonaire, modified after DE BUISONJÉ, 1974.

Fig. 15: Well preserved coral (*Diploria labyrinthiformis*) in the Youngest Pleistocene reef.

Fig. 16: At Seru Grandi in the northeast of Bonaire a well preserved discordance separates an older basis reef complex from a younger one, which is at least of isotope stage 9.

Fig. 17: Well preserved coral *(Montastrea annularis)* in Younger Pleistocene coral reef terrace.

Fig. 18: Slightly uplifted Young Pleistocene coral reef terrace at Washikemba with a well preserved lagoonal area (light brown) to landward.

Fig. 19: Deep incised notch along the east coast near Boka Chikitu **(a)** and the west coast at Devils Mouth **(b)** documenting the sea level high stand of isotope stage 5e in an older reef.

Fig. 20: Topographic map of Bonaire island.

Fig. 21: Artificial salt pans in the south of Bonaire island.

Fig. 22: Notch types depending on the degree of exposure around Bonaire (modified after FOCKE, 1978c).

Fig. 23: Wave distribution and cliff profiles along the Bonaire shorelines (combined from FOCKE, 1978c, and VAN DUYL, 1985).

Fig. 24: Development of a cliff profile along the exposed trade wind shorelines of Bonaire (from SCHEFFERS, 2002a).

Fig. 25: Well developed bench with bioconstructive ledges near Washikemba, east coast.

Fig. 26: The supratidal rock pool zone at exposed coasts of Bonaire.

Fig. 27: Oblique aerial photograph of Salina Term, northwest coast of Bonaire.

Fig. 28: Oblique aerial photograph of Boka Kokolishi, NE Bonaire.

Fig. 29: Boka Kokolishi with algal rims.

Fig. 30: Northern section of Lac Baai with mangroves and coral reefs.

Fig. 31: Oblique aerial photograph of the narrow fringing reefs of Klein Bonaire.

Fig. 32: Mining of tsunami coral debris at Washikemba.

Fig. 33: Coastal infrastructure around Kralendijk, the main town of Bonaire.

Fig. 34: Tracks of hurricanes of category 3-5 for the time period 1871 and 1999. Hurricanes close to Bonaire display names (NOAA, 2004).

Fig. 35: Ten tropical storms and hurricanes up to category 2 came into a distance of 60 nautical miles (= 111 km) of Bonaire from 1851 to 1998. The strong hurricane Ivan passed just north of this radius (STORMCARIB, 2005, NOAA, 2004).

Fig. 36: Track of hurricane *IVAN*, September, 2004.

Fig. 37: Wave crest of hurricane *IVAN* swell along the east coast of Bonaire on September, 9th, 2004, with a height of about 10 m asl.

Fig. 38: Backwash of *IVAN* waves at a coral reef terrace 5-6 m high.

Fig. 39: Dead fish could be found more than 100 m inland at 6 m asl after hurricane *IVAN* waves.

Fig. 40: The spit newly created in 1999 by hurricane *LENNY* north of Slagbaai has been flattened, broadened and curved to inland by swell from hurricane *IVAN*.

Fig. 41: Sand from old bimodal tsunami deposits have been washed out by *IVAN* waves and redistributed on the higher reef flat.

Fig. 42: Along Seru Grandi in NE Bonaire wide sandy flats have been washed over at least till 9 m asl from *IVAN* waves.

Fig. 43: At some places old sand covers have been washed out into the sea, leaving marks of former height along old tsunami boulders.

COASTAL RESPONSE TO EXTREME WAVE EVENTS

Fig. 44: Seaward of the tsunami boulder ramparts a sediment free belt has been created by many strong hurricanes.

Fig. 45: The rock pool belt shows some small destructions but no sediments in the traps after IVAN has passed.

Fig. 46: Along the south coast of Bonaire east of Willemstoren Lighthouse IVAN waves have formed a new broad ridge up to nearly 2 m high, mostly using old tsunami rubble.

Fig. 47: Very good imbrication of the old tsunami ridge is still preserved even in areas of strong IVAN waves.

Fig. 48: Storm waves may alter the morphology of an older tsunami ridge depending on its height above sea level.

Fig. 49: The cliffs along the east coast of Bonaire show break off in parts, and some destruction can be seen in the rock pool belt from IVAN waves.

Fig. 50: One of the larger boulders, broken off the cliff by IVAN waves. Weight is around 6 tons.

Fig. 51: The largest boulder from the sublittoral, weighing 6 tons, now at +4-5 m asl and moved at least 12 m against gravity by IVAN waves.

Fig. 52: A platy fragment of about 20 m² has been broken from the old reef terrace.

Fig. 53: A mushroom rock in Boka Chikitu with a weight of 22 tons has been broken off and dislocated for several meters.

Fig. 54: Old tsunami boulder of more than 20 tons has been tilted at +5 m and more than 100 m distant from the cliff.

Fig. 55: This tsunami boulder with a weight of about 40 tons has been uplifted and settled down on some smaller rocks.

Fig. 56: Delicate new setting of an old tsunami boulder 110 m apart from the sea.

Fig. 57: These two tsunami boulders have been tilted into an upright position by IVAN waves, although weighing 40 to 50 tons.

Fig. 58: This sketch shows the distance of large tsunami boulders from the modern cliff, but moved by hurricane IVAN waves along the Washikemba coast east of Bonaire.

Fig. 59: Distribution of large boulders around the coastlines of Bonaire, moved by waves from hurricane IVAN.

Fig. 60: The unusual track of hurricane LENNY (November, 1999) from the west.

Fig. 61: Fresh hurricane LENNY ridge in Nukowe Bay, NW Bonaire. Darker debris on the left are from a storm event in 1877.

Fig. 62: Fresh storm ridges from LENNY in Nukowe Bay.

Fig. 63: The LENNY ridges show debris tongues to inland, partly burying mangrove bushes.

Fig. 64: North of Slagbaai hurricane LENNY has formed a new rubble spit about 100 m long.

Fig. 65: At Boka Slagbaai houses from 1868 have been destroyed by LENNY waves.

Fig. 66: LENNY waves have filled small incisions in the 5 m high cliff with coral rubble.

Fig. 67: The hard contrast of very dark coral debris with cyanobacterial coating and fresh LENNY ridges can be seen at many places in NW Bonaire.

Fig. 68: Semicircular debris pattern from LENNY waves on top of an old coral reef terrace at about +5 m north of Slagbaai.

Fig. 69: Coral debris is distributed in the coastal vegetation, and some larger boulders have been broken from the cliff by LENNY waves in 1999.

Fig. 70: The sketch shows the areas impacted by hurricane LENNY on Bonaire island.

Fig. 71: A small intermediate terrace of coral debris between the light LENNY terrace of 1999 and the top is the result of a storm from the year 1877. NW coast of Bonaire.

Fig. 72: In Nukowe Bay the storm ridge of 1877 can clearly be distinguished from younger deposits.

Fig. 73: Darker colours on these tonguelike ridge near Goto Meer show an age of many decades for the event as does the vegetation on it. The forms are the result of a storm in 1877. East of Goto Meer.

Fig. 74: The oldest storm deposits of about 900 BP are preserved on top of an older tsunami deposit of about 1300 BP near Salina Tern.

Fig. 75: Old large tsunami boulders (1300 BP) have been overthrown by younger storm debris (from 790 BP) near Salina Tern. See also Fig. 65.

Fig. 76: Typical aspect of the surface of the oldest storm ridge at about 3-3.5 m asl along the NW coast of Bonaire with a lot of small branches from *Acropora cervicornis*.

Fig. 77: The upper half meter of debris on top of the bimodal tsunami sediments near Karpata are from a storm about 900 years ago. See also Fig. 69.

Fig. 78: The base of this sections shows bimodal tsunami deposits (mostly *Acropora cervicornis*), whereas the upper layer is a sand free storm deposit. See also Fig. 77.

Fig. 79: Between Karpata and Goto 3 storm deposits (LENNY of 1999, pre-LENNY of 1877, and a very old one, probably from 790 BP and up to +3.5 m high) can be seen in different sections.

Fig. 80: During former centuries huge *Acropora palmata* must have lived along the NW coast of Bonaire, now incorporated in older tsunami and storm sediments.

Fig. 81: Disturbed and living *Acropora cervicornis* within a sandy environment on a shallow reef flat at the west coast of Bonaire. Water depth approximately 5 m.

Fig. 82: Head corals (*Diploria* sp.) can better survive storm impacts on the shallow fringing reefs along the leeward side of Bonaire. Water depth approximately 5 m.

Fig. 83: Coral debris deposit near Salina Tern: the upper section from a storm about 900 years ago, the lower one, separated by a thin soil layer, shows sand, rubble and boulders from a tsunami event about 1300 years ago.

Fig. 84: Remnant of a tsunami ridge up to 3 m high, some kilometers long and originally 50-80 m wide along the south coast of Bonaire, destroyed by mining along the landward side (left in picture). In the front part fresh erosion by *IVAN* hurricane waves.

Fig. 85: Coral debris in a tsunami ridge about 500 years old from the south coast of Bonaire: fresh broken coral mixed with rounded rubble.

Fig. 86: A wide tsunami rampart, typical for the east coast of Bonaire, at around +5-6 m and up to 400 m inland. Fragments mostly broken off from a cliff and up to more than 200 kg. Age is from about 500 years up to more than 4000 BP.

Fig. 87: Only the base of tsunami ridges and ramparts show the sandy matrix, whereas the upper parts have been washed out by heavy rain and storm splash.

Fig. 88: The tsunami rampart with dark colours on the debris by cyanobacteria etc. as a ridge up to 3.3 m high and 90 m wide in front of Boka Bartol. Young hurricane waves have only washed on a limited amount of smaller debris.

Fig. 89: Aerial photograph (1996) of the Boka Bartol tsunami barrier.

Fig. 90: Aerial photograph of the Salina Tern barrier (1996).

Fig. 91: Fragment of weathered *Acropora palmata* with a weight of about 3 tons as a remnant of a tsunami deposit, excavated by hurricane waves at the northwest coast of Bonaire.

Fig. 92: Some ramparts show imbricated boulders with several tons of weight, here partly excavated from finer debris by hurricane *IVAN* waves. Coast NE of Lac Baai.

Fig. 93: Remnant of a tsunami rampart at Boka Onima, east coast of Bonaire.

Fig. 94: The development of a Holocene tsunami rampart on the Youngest Pleistocene coral reef terrace.

Fig. 95: Mining of tsunami debris and destruction of a rampart along the east coast of Bonaire in the Washikemba area. Aerial photograph of 1996.

Fig. 96: Tsunami boulder at Washikemba, 180 m from the cliff, with very well preserved bioerosive rock pools from the supratidal belt. The boulder has a weight of nearly 5 tons.

Fig. 97: This boulder of about 120 tons and a length of 9.2 m can be found near Playa Chikitu. The extended sandy flats derive from that beach by trade winds and show some small dune ripplets. The seaward limit of the sand has been washed out by strong storm waves.

Fig. 98: A cluster of smaller (1-5 tons) and larger (up to 130 tons) boulders at Seru Grandi between 6 and 8 m asl.

Fig. 99: This sketch shows the position of the mushroom rock (22 tons) broken by hurricane *IVAN* waves on the bottom of Boka Chikitu, and tsunami boulders of 150 and 400 tons at +8 m farther inland.

Fig. 100: Tsunami boulder near Boka Chikitu, weighing 400 tons.

Fig. 101: Even along the leeward side of Bonaire large boulders have been dislocated by tsunami: *Acropora palmata* branches of more than 1.5 tons, or reef rock boulders up to 80 tons and smaller ones at +15 m like at Devil´s Mouth.

Fig. 102: The largest boulder of at least 80 tons at +7 m asl in the Devil´s Mouth area of Bonaire.

Fig. 103: At Spelonk these two pieces of rock, each with a weight of more than 120 tons, have been transported as a single boulder and broken during smash down from a tsunami wave. The site is at +5.5 m asl and 160 m from the cliff.

Fig. 104: Mapping of large boulders from the Spelonk area. At least 89 boulders with a weight of 10 to 50 tons and 32 with more than 50 tons can be found here in a belt 100-150 m from the sea at about +5 m asl.

Fig. 105: Boulder field at Spelonk in dense Conocarpus vegetation.

Fig. 106: Single boulder with 7 m axis, partly covered by vegetation, at Spelonk.

Fig. 107: This large boulder shows its origin from a Pleistocene reef with dominant *Acropora cervicornis*.

Fig. 108: Boulders of more than 100 tons at Seru Grandi. The rock platform at 7-8 m asl is cleared by hurricane waves from smaller debris.

Fig. 109: Another aspect of the Seru Grandi tsunami boulders.

Fig. 110: The largest of the Seru Grandi boulders shows some vegetation on its top, while its environment has been washed out by extreme storm waves.

Fig. 111: Ripple marks in smaller boulder ramparts north of the Seru Grandi rock at +9 m and 100-150 m from the sea (Aerial photograph of 1961).

Fig. 112: This field of ripple marks is more than 200 m wide in the southern Washikemba area and up to 400 m inland at +6-7 m asl. The coarse rampart (boulders 0.5-1 ton) near the cliff shows smaller ripples, as well. Its seaward marked cliff is the result of extreme storm waves, eroding a small part of the original tsunami deposits (1961).

Fig. 113: In undisturbed tsunami ramparts the landward limit of tsunami can be identified by a sharp line, at which the ripple field ends. This is about 200-250 m from the sea (south part of Washikemba) (1961).

Fig. 114: A tsunami rampart from the southern Washikemba area, east coast of Bonaire: the front part has been transformed into a debris cliff by later storms. The coarsest belt of fragments shows ripple marks (in debris of 0.5 to more than 1 m of length and weights of up to 1-1.5 tons!), the landward belt has a lighter colour, and landward bowlike sharp ridges of coarse fragments document the reach of largest tsunami waves. All in all a document for several tsunami or several large waves of a single tsunami (1961).

Fig. 115: This aerial photograph (from 1961) of the Washikemba area shows a pattern of different boulder sizes and colours from the cliff to inland, giving relative indices for age differences of the deposits.

Fig. 116: This aerial picture of 1961 shows at least two different tsunami depositional units by size of debris and colour.

Fig. 117: Aerial picture from 1996 of the mining activities along the coastal stretch of Washikemba.

Fig. 118: Oblique aerial photograph of the Washikemba area with a broad rampart and narrow boulder ridges to inland.

Fig. 119: On Curaçao island along the leeward side tsunami boulder ridges may show several parallel units of different width, sometimes separated by lines of vegetation.

Fig. 120: Trenches in tsunami boulder ridges at Washikemba and Onima may show at least two units of different age, separated by remnants of a brownish soil in between.

Fig. 121: Mapping of tsunami deposits on Bonaire island. Modified from SCHEFFERS (2002a).

Fig. 122: Historical and absolute tsunami data and minimum run up values for the wider Caribbean.

Fig. 123: Absolute data by radiocarbon or ESR for hurricane and tsunami deposits on Bonaire island.

Fig. 124: Summarizing all absolute data of tsunami for Bonaire shows some definite clustering, but still single data distributed in the cluster gaps. The question is, whether these single data are representative for events so far not dated significantly, or whether they represent samples of older age buried on the fringing reefs and dislocated with the next tsunami wave.

Fig. 125: Cut in uplifted Pleistocene coral reef.

Fig. 126: Itinerary and field stops.

List of Tables

Table 1: The SAFFIR-SIMPSON Hurricane Scale.

Table 2: Large boulders moved by waves of hurricane *IVAN*, September, 2004.

Table 3: Historical run up heights in the Caribbean (extracted from LANDER & WHITESIDE, 1997).

ESSENER GEOGRAPHISCHE ARBEITEN

Band 1: Ergebnisse aktueller geographischer Forschungen an der Universität Essen. – 207 Seiten, 47 Abbildungen, 30 Tabellen. 1982. Euro 8.- (ISBN 3-506-72301-4).

Band 2: G. HENKEL (Hg.): Dorfbewohner und Dorfentwicklung. Vorträge und Ergebnisse der Tagung in Bleiwäsche vom 17.-19. März 1982. – 127 Seiten, 6 Abbildungen. 1982. Euro 5.- (ISBN 3-506-72302-3).

Band 3: J.-F. VENZKE: Geoökologische Charakteristik der wüstenhaften Gebiete Islands. – 206 Seiten, 44 Abbildungen, 15 Tabellen, 1 Karte als Beilage. 1982. Euro 9.- (ISBN 3-506-72303-0).

Band 4: J. BIEKER & G. HENKEL: Erhaltung und Erneuerung auf dem Lande. Das Beispiel Hallenberg. – 255 Seiten, 53 Abbildungen, 46 Tabellen, 55 Schwarz-Weiß-Fotos, 2 Farbfotos. 1983. Euro 13.- (ISBN 3-506-72304-9).

Band 5: W. TRAUTMANN: Der kolonialzeitliche Wandel der Kulturlandschaft in Tlaxcala. Ein Beitrag zur historischen Landeskunde Mexikos unter besonderer Berücksichtigung wirtschafts- und sozialgeographischer Aspekte. – 420 Seiten, 13 Abbildungen, 25 Fotos, 10 Karten, 7 Tabellen. 1983. Euro 13.- (ISBN 3-506-72305-7).

Band 6: D. KELLETAT (Hg.): Beiträge zum 1. Essener Symposium zur Küstenforschung. – 312 Seiten, 97 Abbildungen, 7 Tabellen. 1983. Euro 13.- (ISBN 3-506-72306-5).

Band 7: D. KELLETAT: Internationale Bibliographie zur regionalen und allgemeinen Küstenmorphologie (ab 1960). – 218 Seiten, 1 Abbildung. 1983. Euro 8.- (ISBN 3-506-72307-3).

Band 8: G. HENKEL & H.-J. NITZ (Hg.): Ländliche Siedlungen einheimischer Völker Außereuropas – Genetische Schichtung und gegenwärtige Entwicklungsprozesse. Arbeitskreissitzung des 44. Deutschen Geographentages. – 148 Seiten, 35 Abbildungen, 21 Fotos, 4 Karten, 5 Tabellen. 1984. Euro 8.- (ISBN 3-506-72308-1).

Band 9: H.-W. WEHLING: Wohnstandorte und Wohnumfeldprobleme in der Kernzone des Ruhrgebietes. – 285 Seiten, 38 Abbildungen, 24 Tabellen, 10 Übersichten. 1984. Euro 13.- (ISBN 3-506-72309-X).

Band 10: D. KELLETAT (Hg.): Beiträge zur Geomorphologie der Varanger-Halbinsel, Nord-Norwegen (KELLETAT: Studien zur spät- und postglazialen Küstenentwicklung der Varanger-Halbinsel, Nord-Norwegen; MEIER: Studien zur Verbreitung, Morphologie, Morphodynamik und Ökologie von Palsas auf der Varanger-Halbinsel, Nord-Norwegen). – 243 Seiten, 63 Abbildungen, 8 Abbildungen, 45 Tabellen. 1985. Euro 13.- (ISBN 3-506-72310-3).

Band 11: D. KELLETAT: Internationale Bibliographie zur regionalen und allgemeinen Küstenmorphologie (ab 1960) – 1. Supplement (1960-1985). – 244 Seiten, 1 Abbildung. 1985. Euro 8.- (ISBN 3-506-72311-1).

Band 12: H.-W. WEHLING: Das Nutzungsgefüge der Essener Innenstadt. – Bestand und Veränderungen seit 1978. – 1986 (ISBN 3-506-72312-X).

Band 13: W. KREUER: Landschaftsbewertung und Erholungsverkehr im Reichswald bei Kleve. Eine Studie zur Praxis der Erholungsplanung. – 205 Seiten, 45 Abbildungen, 46 Tabellen. 1986. Euro 10.- (ISBN 3-506-72313-8).

Band 14: D. KELLETAT & H.-W. WEHLING: Beiträge zur Geographie Nord-Schottlands (KELLETAT: Die Bedeutung biogener Formung im Felslitoral Nord-Schottlands. WEHLING: Leben am Rande Europas. Wirtschafts- und Sozialstrukturen in der Crofting-Gemeinde Durness). – 176 Seiten, 85 Abbildungen, 6 Tabellen. 1986. Euro 10.- (ISBN 3-506-72314-6).

Band 15: G. HENKEL (Hg.): Kommunale Gebietsreform und Autonomie im ländlichen Raum. Vorträge und Ergebnisse der Tagung in Bleiwäsche vom 12.-13. Mai 1986. – 160 Seiten. 1986. Euro 8.- (ISBN 3-506-72315-4).

Band 16: G. HENKEL (Hg.): Kultur auf dem Lande. Vorträge und Ergebnisse des 6. Dorfsymposiums in Bleiwäsche vom 16.-17. Mai 1988. – 231 Seiten, 22 Fotos, 2 Tabellen. 1988. Euro 11.- (ISBN 3-506-72316-4).

Band 17: D. KELLETAT (Hg.): Neue Ergebnisse zur Küstenforschung. Vorträge der Jahrestagung Wilhelmshaven 18.-19. Mai 1989. – 388 Seiten, 23 Fotos, 119 Abbildungen. 1989. Euro 13.- (ISBN 3-506-72317-0).

Band 18: E. C. F. BIRD & D. KELLETAT (Eds.): Zonality of Coastal Geomorphology and Ecology. Proceedings of the Sylt Symposium, August 30 - September 3, 1989. – 295 Seiten, 88 Fotos, 52 Abbildungen, 7 Tabellen. 1989. Euro 15.- (ISBN 3-506-72318-9).

Band 19: G. HENKEL & R. TIGGEMANN (Hg.): Kommunale Gebietsreform – Bilanzen und Bewertungen. Beiträge und Ergebnisse der Fachsitzung des 47. Deutschen Geographentages Saarbrücken 1989. – 124 Seiten, 9 Abbildungen, 5 Tabellen. 1990. Euro 8.- (ISBN 3-506-72319-7).

Band 20: W. KREUER: Tagebuch der Heilig Land-Reise des Grafen Gaudenz von Kirchberg, Vogt von Matsch/Südtirol im Jahre 1470. – 349 Seiten, 3 beigelegte Karten, 7 Karten, 26 Abbildungen. 1990. Euro 23.- (ISBN 3-506-72320-0).

Band 21: J.-F. VENZKE: Beiträge zur Geoökologie der borealen Landschaftszone. Geländeklimatologische und pedologische Studien in Nord-Schweden. – 254 Seiten, 27 Fotos, 81 Abbildungen, 12 Tabellen. 1990. Euro 15.- (ISBN 3-506-72321-9).

Band 22: G. HENKEL (Hg.): Schadet die Wissenschaft dem Dorf? Vorträge und Ergebnisse des 7. Dorfsymposiums in Bleiwäsche vom 7.- 8. Mai 1990. – 150 Seiten. 1990. Euro 9.- (ISBN 3-506-72322-5).

Band 23: D. KELLETAT & L. ZIMMERMANN: Verbreitung und Formtypen rezenter und subrezenter organischer Gesteinsbildungen an den Küsten Kretas. – 163 Seiten, 37 Fotos, 45 Abbildungen, 7 Tabellen. 1991. Euro 14.- (ISBN 3-506-72323-5).

Band 24: G. HENKEL (Hg.): Der ländliche Raum in den neuen Bundesländern. Vorträge und Ergebnisse des 8. Essener Dorfsymposiums in Wilhelmsthal. Gemeinde Eckardtshausen in Thüringen (bei Eisenach) vom 25.-26. Mai 1992. – 105 Seiten. 1992. Euro 11.- (ISBN 3-506-72324-3).

Band 25: J.-F. VENZKE (Hg.): Zur Ökologie und Gefährdung der borealen Landschaftszone. Beiträge zu einem borealgeographischen Kolloquium an der Universität Essen im Wintersemester 1993/94. – 173 Seiten. 1994. Euro 21.- (ISBN 3-506-72325-1).

Band 26: G. HENKEL (Hg.): Außerlandwirtschaftliche Arbeitsplätze im ländlichen Raum. Vorträge und Ergebnisse des 9. Essener Dorfsymposiums in Bleiwäsche vom 9.-10.5.1994. – 154 Seiten. 1995. Euro 35.- (ISBN 3-88474-291-4).

Band 27: H.-W. WEHLING: City im Wandel. Die Nutzungsstruktur der Essener Innenstadt 1995. – 155 Seiten. 40 Abbildungen, 4 Faltkarten, 471 Tabellen. 1996. Euro 76.- (ISBN 3-88474-537-9). *vergriffen*

Band 28: G. HENKEL (Hg.): Das Dorf in Wissenschaft und Kunst – Vorträge und Ergebnisse des 10. Dorfsymposiums in Bleiwäsche vom 13. und 14. Mai 1996. – 106 Seiten, 35 Abbildungen. 1997. Euro 30.- (ISBN 3-88474-579-4). *vergriffen*

Band 29: G. SCHELLMANN: Jungkänozoische Landschaftsgeschichte Patagoniens (Argentinien). Andine Vorlandvergletscherungen, Talentwicklung und marine Terrassen. – 216 Seiten, 102 Abbildungen, 23 Tabellen, 36 Bilder, 7 Tabellenseiten im Anhang. 1998. Euro 81.- (ISBN 3-88474-671-5). *vergriffen*

Band 30: G. HENKEL (Hg.): 20 Jahre Dorferneuerung – Bilanzen und Perspektiven für die Zukunft. Vorträge des 11. Dorfsymposiums in Bleiwäsche vom 25.-26. Mai 1998. – 128 Seiten, 21 Abbildungen, 8 Tabellen. 1999. Euro 20.- (ISBN 3-9803484-5-8).

Band 31: G. HENKEL (Hg.): Das Dorf im Einflussbereich von Großstädten. – 134 Seiten, 43 Abbildungen, 15 Tabellen. 2000. Euro 20.- (ISBN 3-9803484-7-4). *vergriffen*

Band 32: D. KELLETAT & G. SCHELLMANN (Hg.): Küstenforschung auf Zypern. Tsunamiereignisse und chrono-stratigraphische Untersuchungen. – 104 Seiten, 118 Abbildungen, 2 Tabellen. 2001. Euro 20.- (ISBN 3-9803484-8-2).

Band 33: A. SCHEFFERS: Paleotsunamis in the Caribbean. Field Evidences and Datings from Aruba, Curaçao and Bonaire. – 185 Seiten, 186 Abbildungen (davon 12 in Farbe), 18 Tabellen. 2002. Euro 34.- (ISBN 3-9803484-9-0).

Band 34: G. HENKEL (Hg.): Bürgerbüro, Bürgerladen KOMM-IN. Multifunktionale Dienstleistungszentren im ländlichen Raum – Vorträge und Ergebnisse des 13. Dorfsymposiums in Bleiwäsche vom 5.-7. Mai 2002. – 105 Seiten, 43 Abbildungen, 8 Tabellen. 2002. Euro 20.- (ISBN 3-9808567-0-4).

Band 35: D. KELLETAT (Hg.): Neue Ergebnisse der Küsten- und Meeresforschung. – 21. Jahrestagung des Arbeitskreises "Geographie der Meere und Küsten", 1.-2. Mai 2003 in Essen. – 205 Seiten, 130 Abbildungen, 27 Tabellen. 2003. Euro 26.- (ISBN 3-9808567-1-2).

Band 36: G. HENKEL (Hg.): Dörfliche Lebensstile – Mythos, Chance oder Hemmschuh der ländlichen Entwicklung ? – Vorträge und Ergebnisse des 14. Dorfsymposiums in Bleiwäsche vom 16.-18. Mai 2004. – 139 Seiten, 19 Abbildungen, 10 Tabellen. 2004. Euro 22.- (ISBN 3-9808567-2-0).

Band 37: A. SCHEFFERS: Coastal Response to Extreme Wave Events – Hurricanes and Tsunami on Bonaire. – Fieldguide zum "First International Tsunami Field Symposium" im März 2006 auf Bonaire, 100 Seiten, 126 Abbildungen, 3 Tabellen. 2005. Euro 32.- (ISBN 3-9808567-3-9).

Zu beziehen durch:

Band 1 - 25, ab Band 30 und Sonderbände:

Selbstverlag – Institut für Geographie
Universität Duisburg-Essen, Fachbereich Biologie und Geographie, Universitätsstr. 15, D-45117 Essen,
e-mail: geographie@uni-essen.de, Fax: 0201 / 1832818

Band 26 - 29:

Klartext Verlag
Dickmannstr. 2, D-45143 Essen, Tel.: 0201 / 86206-31/32, Fax: 0201 / 86206-22

ESSENER GEOGRAPHISCHE SCHRIFTEN

Band 1: F. SCHULTE-DERNE & H.-W. WEHLING: Atlas des Handwerks in Gelsenkirchen. – 128 Seiten, mit 10 Abbildungen, 53 Tabellen, 22 Farbfotos, 54 Farbkarten. Essen 1993. Euro 25.- (ISBN 3-9803484-0-7).

Band 2: W. KREUER: Imago Civitatis. Stadtbildsprache des Spätmittelalters. – 195 Seiten, mit 71 Abbildungen, 32 Faksimiles, Großformat im Schuber. Essen 1993. Euro 250.- / gefalzt: Euro 200.- (ISBN 3-9803484-1-5).

Band 3: W. KREUER.: Monumenta Cartographica. - 63 Seiten, 26 Abbildungen, 6 Kartentafeln, Großformat in Leinenkassette. Essen 1996. Euro 150.- (ISBN 3-9803484-3-1)

Zu beziehen durch:

Selbstverlag – Institut für Geographie
Universität Duisburg-Essen, Fachbereich Biologie und Geographie, Universitätsstr. 15, D-45117 Essen,
e-mail: geographie@uni-essen.de, Fax: 0201 / 1832811

Sonstige Publikationen

D. KELLETAT (Ed.): Field Methods and Models to Quantify Rapid Coastal Changes – Crete Field Symposium 1994, Program, Abstracts and Field Guide. – 80 Seiten, 27 Abbildungen und Karten, 3 Tabellen. Essen 1994. Euro 10.- (ISBN 3-9803484-2-3).

K. FEHN & H.W. WEHLING (Hg.): Bergbau und Industrielandschaften. – 327 Seiten, 79 Abbildungen und Karten, 8 Tabellen. Essen 1999. Euro 20.- (ISBN 3-9803484-6-6).

– ebenfalls über den Selbstverlag zu beziehen –